D1426655

HYDRAULIC DESIGN OF STEPPED CASCADES, CHANNELS, WEIRS AND SPILLWAYS

Elsevier Titles of Related Interest

YALIN
River Mechanics

TANAKA & CRUSE
Boundary Element Methods in Applied Mechanics

USCOLD (US Committee on Large Dams)
Development of Dam Engineering in the United States

JAPAN SOCIETY OF MECHANICAL ENGINEERS
Visualized Flow

TANIDA
Atlas of Visualization

WILLIAMS & ELDER
Fluid Physics for Oceanographers and Physicists

Related Journals
(free specimen copy gladly sent on request)

Advances in Engineering Software
Advances in Water Resources
Applied Ocean Research
Coastal Engineering
Computers and Fluids
Dynamics of Atmospheres and Oceans
Engineering Analysis with Boundary Elements
Finite Elements in Analysis and Design
Fluid Abstracts: Process Engineering
Fluid Dynamics Research
International Journal of Engineering Science
Journal of Non-Newtonian Fluid Mechanics
Journal of Hydrology
Minerals Engineering
Ocean Engineering
Marine Structures
Marine Geology
Water Research
Wave Motion

HYDRAULIC DESIGN OF STEPPED CASCADES, CHANNELS, WEIRS AND SPILLWAYS

by
Hubert Chanson

Hydraulics and Environmental Engineering,
Department of Civil Engineering, The University of Queensland,
Brisbane QLD 4072, Australia

Pergamon

U.K.   Elsevier Science Ltd, The Boulevard, Langford Lane, Kidlington, Oxford,
            OX5 1GB, England

U.S.A.   Elsevier Science Inc., 660 White Plains Road, Tarrytown, New York
            10591-5153, U.S.A.

JAPAN   Elsevier Science Japan, Tsunashima Building Annex, 3-20-12 Yushima,
            Bunkyo-ku, Tokyo 113, Japan

First edition 1994

Library of Congress Cataloging in Publication Data
A catalogue record for this book is available from the Library of Congress

British Library Cataloguing in Publication Data
A catalogue record for this book is available from the British Library

ISBN 0 08 041918 6

*In order to make this volume available as economically and as rapidly as
possible the author's typescript has been reproduced in its original form. This
method unfortunately has its typographical limitations but it is hoped that they
in no way distract the reader.*

Printed and bound in Great Britain by Redwood Books, Trowbridge

A mes parents,
A ma famille.

TABLE OF CONTENTS

Abstract

Stepped channels and spillways are used since more than 2,500 years. Recently, new construction materials (e.g. RCC, gabions) have increased the interest for stepped chutes. The steps increase significantly the rate of energy dissipation taking place along the chute and reduce the size of the required downstream energy dissipation basin. Stepped cascades are used also in water treatment plants to enhance the air-water transfer of atmospheric gases (e.g. oxygen, nitrogen) and of volatile organic components (VOC).

A first part describes the historical progress of stepped channels and spillways from the Antiquity up to today. Then the monograph reviews the hydraulic characteristics of stepped channel flows. Two different flow regimes can take place : nappe flow regime for small discharges and flat channel slopes, and skimming flow regime. The hydraulics of each flow regime is described. The effects of flow aeration and air bubble entrainment are discussed. Further the process of air-water gas transfer taking place above stepped chute is described. Later practical examples of hydraulic design are presented : e.g. stepped fountains, stepped weirs, gabion stepped spillways, earth dam spillways with precast concrete blocks, roller compacted concrete (RCC) weirs, debris dams. At the end, the author presents a critical review of the risks of accidents and failures with stepped channels. It is shown that the hydrodynamic forces on the step faces are much larger than on smooth chute bottoms.

Résumé

Les canaux et évacuateurs de crues en marches d'escaliers sont utilisés depuis plus de 2500 ans. Récemment, l'introduction de nouveaux matériaux de constructions (ex.: BCR, gabions) a accru l'intérêt pour des évacuateurs de crues en marches d'escalier. Les marches d'escalier augmentent considérablement la dissipation d'énergie au long du coursier, et permettent de réduire la taille des bassins de dissipation avals. Les cascades en marches d'escalier sont utilisées aussi dans les stations de traitement d'eau, pour accroitre le transfert de gaz dissouts entre l'air et l'eau : ceci peut s'appliquer aux gaz atmosphériques (oxygène, azote) ou aux composés organiques volatiles.

Dnas une première partie, l'auteur décrit les progrès et développements de canaux en marches d'escalier, depuis l'Antiquité. Puis cet ouvrage passe en revue les caractéristiques des écoulements sur des coursiers en marches d'escalier. On met en évidence deux types d'écoulement : un écoulement en nappe pour des petits débits et faibles pentes, et un écoulement extrémement turbulent. Les caractéristiques hydrauliques de chaque régime sont décrites. On présente et discute les effets de l'entraînement d'air sur ces écoulements. Puis on développe les calculs de transferts de gaz atmosphériques et volatiles. Ensuite, l'auteur présente des cas d'étude qui sont détaillés un par un : fontaines en marches d'escalier, évacuateurs de crues à gabions en marches d'escalier, canaux sur remblai renforcés avec des revêtements en blocs de béton pré-fabriqués, évacuateurs de crues pour des barrages en béton compacté au rouleau, barrages de contrôle des torrents de débris. Enfin, on présente une étude critique des risques d'accidents avec des canaux en marches d'escalier. On montre que les efforts hydrodynamiques sur les marches d'escalier sont beaucoup plus important que sur les coursiers de chutes lisses.

List of Symbols

A	gas-liquid interface area (m^2);
A_G	constant (defined by GAMESON et al. 1958);
A_w	cross-section area (m^2) of the water flow;
A_1, A_2	constants;
A_3, A_4	constants;
a	specific interface area (m^{-1}) defined as the air-water surface area per unit volume of air and water;
a_{mean}	mean interface area (m^{-1}) in a cross-section of fluid normal to the flow direction :

$$a_{mean} = \frac{1}{Y_{90}} * \int_0^{Y_{90}} a * dy$$

a_1, a_2	constants;
B	constant (defined by the Department of Environment 1973);
B'	integration constant (defined by WOOD 1984);
B_1, B_2, B_3	constants;
C	air concentration defined as the volume of air per unit volume; (Note : it is also called void fraction);
C_{DS}	downstream dissolved gas concentration (kg/m^3);
C_{US}	upstream dissolved gas concentration (kg/m^3);
C_b	air concentration next to the chute invert : i.e. at the outer edge of the air concentration boundary layer;
C_e	equilibrium depth averaged air concentration for uniform flow;
C_{gas}	concentration of dissolved gas in water (kg/m^3);
C_{max}	maximum air concentration in the fully-developed flow region of a plunging jet;
$(C_{max})_o$	initial maximum air content of a plunging jet;
C_{mean}	depth averaged air concentration defined as : $(1 - Y_{90}) * C_{mean} = d$;
C_s	gas saturation concentration in water (kg/m^3);
Chl	chlorinity;
D	molecular diffusivity (m^2/s);
D_{gas}	molecular diffusivity of gas in water (m^2/s);
D_H	hydraulic diameter (m) defined as : $D_H = 4 * \frac{A_w}{P_w}$; $\left(D_H = \frac{4 * d*W}{W + 2*d} \text{ for a rectangular channel}\right)$
$(D_H)_i$	hydraulic diameter (m) of a jet impacting a pool of water;
D_p	penetration depth (m) of air bubbles entrained by plunging jet;
d	1- flow depth measured normal to the channel slope at the edge of a step;

2- characteristic depth (m) defined as : $d = \int_{C=0\%}^{C=90\%} (1 - C)\,{}^*dy$;

3- jet thickness or jet diameter (m);

d_I flow depth at the inception point (m);

d_{ab} air bubble diameter (m);

$(d_{ab})_{max}$ maximum bubble size (m) in the shear flow region of a hydraulic jump;

$(d_{ab})_{mean}$ mean bubble size (m) in the shear flow region of a hydraulic jump;

d_b flow depth at the brink of a step (m);

d_c critical flow depth (m); for a rectangular channel : $d_c = \sqrt[3]{q_w^2/g}$;

$(d_c)_{onset}$ critical flow depth (m) at the onset of skimming flow regime;

d_i jet thickness (m) at the impact of the nappe with the receiving pool in nappe flow

             regime;

d_m maximum air bubble size (m) in turbulent shear flow;

$(d_m)_*$ maximum air bubble size (m) that would split in turbulent shear layer flow

d_o uniform flow depth (m);

d_p flow depth in the pool beneath the nappe (m);

d_t 1- tailwater depth (m);

             2- pool height (m);

d_{wp} water droplet diameter (m);

d_1 flow depth (m) upstream of a hydraulic jump;

d_2 flow depth (m) downstream of a hydraulic jump;

d_{50} median boulder diameter (m);

E   aeration efficiency defined as : $E = (C_{DS} - C_{US})/(C_s - C_{US})$;

E_C kinetic energy correction coefficient (Coriolis coefficient);

E_i aeration efficiency of the i-th step;

F   nappe oscillation frequency (Hz);

Fr   Froude number;

Fr_b Froude number defined in term of the brink depth : $Fr_b = q_w/\sqrt{g * d_b^3}$

Fr_e Froude number at the onset of air entrainment : $Fr_e = V_e/\sqrt{g * d}$;

Fr_1 Froude number upstream of a hydraulic jump : $Fr_1 = q_w/\sqrt{g * d_1^3}$

Fr_2 Froude number downstream of a hydraulic jump : $Fr_2 = q_w/\sqrt{g * d_2^3}$

F_* Froude number defined in term of the roughness height :

             1- $F_* = q_w/\sqrt{g * \sin\alpha * (k_s')^3}$ for smooth chute;

             2- $F_* = q_w/\sqrt{g * \sin\alpha * k_s^3}$ for skimming flow;

f   friction factor for non-aerated flow;

f_e friction factor for aerated flow;

G'   integration constant (defined by WOOD 1984);

g	gravity constant (m/s^2) or acceleration of gravity;
H	total head (m);
H_{dam}	dam height (m) : i.e. dam crest head above downstream toe;
H_{gas}	Henry's law constant
H_{max}	maximum head available (m);
H_o	free-surface elevation (m) above the spillway crest;
H_{res}	residual head at the bottom of the spillway (m);
H_1	total head (m) upstream of a hydraulic jump;
H_2	total head (m) downstream of a hydraulic jump;
h	height of steps (m) (measured vertically);
I	1- integer;
	2- number of wavelengths in a free-falling nappe;
i	integer;
K	Von Karman constant;
K_e	entrainment rate coefficient (defined by CHANSON 1993a);
K_L	liquid film coefficient (m/s);
K_M	coefficient of mass transfer (m/s);
K', K'', K_1	constants;
K_2, K_3, K_4	constants;
k'	constant (defined by AVERY and NOVAK 1978);
k_{AN}	constant (defined by AVERY and NOVAK 1978);
k_{N1}, k_{N2}	constants (defined by NAKASONE 1987);
k_{N3}	constant (defined by NAKASONE 1987);
k_s	1 roughness height (m) measured perpendicular to the channel bed;
	2- step dimension (m) measured normal to the flow direction : $k_s = h * \cos\alpha$;
k_s'	surface (skin) roughness height (m);
	Note : for a smooth channel : $k_s' = k_s$;
k_1, k_2	constants;
k_3, k_4	constants;
L	distance along the spillway (m);
L_a	1- aeration length (m) of a hydraulic jump;
	2- bubble zone length (m) of a drop structure measured from the vertical face of the drop;
L_d	length (m) of the drop measured from the vertical face of the step;
L_I	distance from the start of growth of boundary layer to the inception point of air entrainment;
L_r	roller length (m) of a fully-developed hydraulic jump;
L_s	roughness spacing (m);
$L_{spillway}$	spillway length (m);

L_t measure of the turbulent length scale (m);

l   horizontal length of steps (m) (measured perpendicular to the vertical direction);

l_s roughness length (m);

Ln   Neperian logarithm;

Log_{10} decimal logarithm : $Log_{10}(X) = Ln(X)/Ln(10)$;

M_{gas} mass of dissolved gas (kg);

N   exponent of the velocity power law;

N_{step} number of steps;

n   constant;

n_{gabion} gabion porosity (or void fraction);

$n_{Manning}$ Manning coefficient ($s/m^{1/3}$);

P   pressure (Pa);

P_N pressure gradient number : $P_N = \Delta P/(\rho_w * g * d)$;

P_{Nb} pressure gradient number defined in term of the brink depth: $P_{Nb} = \Delta P/(\rho_w * g * d_b)$;

P_{atm} atmospheric pressure (Pa);

P_{gas} partial pressure of gas in air (Pa);

P_{hyd} hydrostatic pressure (Pa) : $P_{hyd} = \rho_w * g * d * \cos\alpha$;

P_{res} residual hydraulic power (W) : $P_{res} = \rho_w * g * Q_w * H_{res}$;

P_s mean stagnation pressure (Pa) at the impact of a falling nappe;

Pstd   standard pressure : Pstd = 1 atm = 1.01325 E+5 Pa;

P_v vapour pressure (Pa);

P_w wetted perimeter (m);

Q   discharge (m^3/s);

Q_{air}^{jet} quantity of air entrained (m^3/s) by plunging jet;

Q_{air}^{HJ} quantity of air entrained (m^3/s) by hydraulic jump;

Q_{air}^{nappe} nappe ventilation air discharge (m^3/s);

Q_w' incoming discharge (m^3/s) into a side channel;

q   discharge per unit width (m^2/s);

q_w' incoming discharge per unit width (m^2/s) into a side channel;

$(q_w)_c$ characteristic discharge per unit width (m^2/s) for which the growing boundary layer reaches the free-surface at the spillway end;

R   radius (m);

Re   Reynolds number defined as :

      1- $Re = \rho_w * \dfrac{V * d}{\mu_w}$, or

      2- $Re = \rho_w * \dfrac{U_w * D_H}{\mu_w}$;

r   deficit ratio : $r = (C_s - C_{US})/(C_s - C_{DS})$;

r_T deficit ratio at temperature T;

r_{15}	deficit ratio at 15 Celsius;
r_{20}	deficit ratio at 20 Celsius;
r_{25}	deficit ratio at 25 Celsius;
Sal	salinity;
s	curvilinear coordinate (m) : i.e., distance measured along the channel bottom;
T	temperature (K);
TC	temperature expressed in Celsius : TC = TK - 273.16;
Teta	coefficient;
TK	temperature expressed in Kelvin : TK = TC + 273.16;
T_o	reference temperature (K);
Tu	turbulence intensity defined as : Tu = u'/V;
t	time (s);
t', t"	time (s);
U_w	flow velocity (m/s) : U_w = q_w/d ;
$(U_w)_o$	mean flow velocity (m) of uniform flow;
u'	root mean square of longitudinal component of turbulent velocity (m/s);
u_r	rise bubble velocity (m/s);
u_t	measure of the turbulent velocity fluctuation (m/s);
V	velocity (m/s);
V_I	free-surface velocity (m/s) at the inception point of air entrainment;
V_b	flow velocity (m/s) at the brink of a step;
V_c	critical velocity (m/s); for a rectangular channel : $V_c = \sqrt[3]{g * q_w}$;
V_e	onset velocity (m/s) for air entrainment;
V_i	velocity (m/s) of the falling nappe at the intersection of the nappe and the receiving pool in nappe flow regime;
V_{max}	maximum velocity (m/s) near the free-surface;
V_o	initial jet velocity (m/s);
V_x	velocity component m/s in the x-direction;
V_y	velocity component m/s in the y-direction;
V_*	shear velocity (m/s);
V_1	mean flow velocity (m/s) upstream of a hydraulic jump : V_1 = q_w/d_1;
V_{90}	characteristic velocity (m/s) where the air concentration is 90%;
v'	root mean square of lateral component of turbulent velocity (m/s);
v'^2	spatial average (m/s) of the square of the velocity differences over a distance equal to the bubble diameter;
W	channel width (m);
We	Weber number;
$(We)_c$	critical Weber number for bubble break-up in turbulent shear flow;

$(We)_e$ self-aerated flow Weber number : $(We)_e = \rho_w * \dfrac{V_{90}{}^2 * Y_{90}}{\sigma}$;

W' width (m);

W_1 width (m) of side channel (fig. 7-1);

x longitudinal distance (m);

x' dimensionless distance;

x_1 horizontal distance (fig. A-2);

Y_{Cmax} distance (m) normal to the jet centreline where $C = C_{max}$;

$Y_{0.1}$ distance (m) normal to the jet centreline where $C = 0.1*C_{max}$;

Y_{90} characteristic depth (m) where the air concentration is 90%;

$(Y_{90})_o$ characteristic depth (m) where the air concentration is 90% of uniform flow;

y 1- distance (m) from the bottom measured perpendicular to the spillway surface;

            2- distance (m) from the pseudo-bottom (formed by the step edges) measured perpendicular to the flow direction;

y' dimensionless distance :

            1- $y' = y/Y_{90}$;

            2- $y' = y_1/d_b$;

y_1 vertical distance (fig. A-2);

y_{50} distance (m) normal to the jet centreline where $V = V_o/2$;

Z Morton number defined as : $Z = \dfrac{g * \mu_w{}^4}{\rho_w * \sigma^3}$;

z bed elevation (or altitude) positive upwards (m);

α channel slope;

α' spillway slope;

$\alpha_1, \alpha_2, \alpha_3$ constants;

ΔH 1- head loss (m);

            2- fall height (m);

$\Delta(H_{res})$ increase of residual energy (m) caused by the flow aeration;

ΔP subpressure (Pa) in the cavity beneath the nappe;

ΔV difference between the upstream and downstream velocities (m/s);

δ step slope : i.e. angle between the step and the horizontal :

            $\delta > 0$ inclined upward step;

            $\delta < 0$ inclined downward step;

δ_{ab} air concentration boundary layer thickness (m);

Γ_y vertical acceleration (m/s^2) of a free-falling nappe;

Λ constant

μ dynamic viscosity (N.s/m^2);

μ_T dynamic viscosity (N.s/m^2) at temperature T;

ν	kinematic viscosity (m^2/s);
ν_T	kinematic viscosity (m^2/s) at temperature T;
ν_t	turbulent kinematic viscosity (m^2/s);
Θ	constant;
θ	angle of the impinging jet and the horizontal step;
θ_b	initial angle of the streamlines at the brink of a step;
θ_1	inner spread angle of a plunging jet in the developing flow region;
θ_2	outer spread angle of a plunging jet in the developing flow region;
θ_3	outer spread angle of a plunging jet in the full-developed flow region;
ρ	density (kg/m^3);
ρ_{gas}	gas density (kg/m^3);
ρ_s	density of rockfill (kg/m^3);
ρ_T	density (kg/m^3) at temperature T;
σ	surface tension between air and water (N/m);
σ_T	surface tension between air and water (N/m) at temperature T;
σ_c	compressive strength (Pa);
τ_o	average bottom shear stress (Pa);
ω	vorticity (rad/s);
ξ	dimensionless parameter;
$\varnothing$	diameter (m);
$\varnothing_s$	circular strip roughness diameter (m);

Subscript

air	air flow;
b	flow conditions at the brink of a step;
c	critical flow conditions;
char	transition from nappe flow with fully-developed hydraulic jump to nappe flow with partially developed jump;
I	inception point of free-surface aeration;
o	uniform flow;
w	water flow;
15, 20, 25	at 15 Celsius, 20 Celsius, 25 Celsius;

Superscript

HJ	hydraulic jump;
jet	plunging jet;
nappe	nappe ventilation.

Glossary

Abutment : part of the valley side against which the dam is constructed. Artificial abutments are sometimes constructed to take the thrust of an arch where there is no suitable natural abutment.

Air concentration : concentration of undissolved air defined as the volume of air per unit volume of air and water.

Arch dam : dam in plan dependent on arch action for its strength.

Arched dam : gravity dam which is curved in pan. Alternatives include "curved-gravity dam" and "arch-gravity dam".

Assyria : land to the North of Babylon comprising, in its greatest extent, a territory between the Euphrates and the mountain slopes East of the Tigris. The Assyrian Kingdom lasted from about B.C. 2300 to B.C. 606.

BAKHMETEFF : Boris Alexandrovitch BAKHMETEFF (1880-1951) was a Russian Hydraulician. In 19112, he developed the concept of specific energy and energy diagram for open channel flows.

Barrage : French word for dam or weir but commonly used to described large dam structure in English.

BAZIN : H. BAZIN was a French hydraulician (1829-1917), engineer and member of the French Corps des Ponts-et-Chaussées. He worked as an assistant of Henri P.G. DARCY.

BERNOULLI : Daniel BERNOULLI was a Swiss mathematician (1700-1782) who developed the Bernoulli equation in his 'Hydrodynamica' textbook (1st draft in 1733).

BIDONE : Giorgio BIDONE (1781-1839) was an Italian Hydraulician. His experimental investigations on the hydraulic jump wee published in 1820 and 1826.

Boundary layer : thin layer of fluid in the neighbourhood of a solid boundary where friction plays an essential part (i.e. flow region affected by the presence of the boundary). A range of velocities exists across the boundary layer from zero at the boundary to the free-stream velocity at the outer edge of the boundary layer.

BOUSSINESQ : J. BOUSSINESQ was a French hydrodynamicist (1842-1929). His treatise "Essai sur la théorie des eaux courantes" (1877) remains an outstanding contribution in hydraulics literature.

Boussinesq coefficient : momentum correction coefficient named after J. BOUSSINESQ, French Mathematician, who first proposed it (BOUSSINESQ 1877).

Braccio : ancient measure of length (from the Italian 'braccia'). One braccio equals 0.6096 m (or 2 ft).

Buttress dam : a special type of dam in which the water face consists of a series of slabs or arches supported on their air faces by a series of buttresses.

Cavitation : formation of vapour bubbles within a liquid at low-pressure regions that occur in places where the liquid has been accelerated (e.g. turbines, marine propellers). Cavitation produces damaging erosion, additional noise, vibrations and dissipate some flow energy.

Chimu : Indian of a Yuncan tribe dwelling near Trujillo on the North-West coast of Peru. The Chimu empire lasted from A.D. 1250 to 1466. It was overrun by the Incas in 1466.

Cofferdam : temporary structure enclosing all or part of the construction area so that construction can proceed in dry. A diversion cofferdam diverts a stream into a pipe or channel.

CORIOLIS : Gustave C. CORIOLIS (1792-1843) was a French engineer and mathematician who first described the Coriolis force (i.e. effect of motion on a rotating body).

Coriolis coefficient : kinetic energy correction coefficient named after G.C. CORIOLIS who introduced first the correction coefficient (CORIOLIS 1836).

Creager profile : spillway shape developed from a mathematical extension of the original data of BAZIN in 1886-88 (CREAGER 1917).

Crest of spillway : upper part of a spillway. The term "crest of dam" refers to the upper part of an uncontrolled overflow.

Crib : 1- framework of bars or spars for strengthening. 2- frame of logs or beams to be filled with stones, rubble or filling material and sunk as a foundation or retaining wall.

Crib dam : gravity dam built up of boxes, cribs, crossed timbers or gabions, and filled with earth or rock.

Culvert : covered channel of relatively short length installed to drain water through an embankment (e.g. highway, railroad, dam).

Cyclopean dam : gravity masonry dam made of very large stones embedded in concrete.

DARCY : Henri P.G. DARCY (1805-1858) was a French civil engineer. He gave his name to the Darcy-Weisbach friction factor.

Debris : Debris comprise mainly large boulders, rock fragments, gravel-sized to clay-sized material, tree and wood material that accumulate in creeks.

Diversion channel : waterway used to divert water from its natural course.

Diversion dam : dam or weir built across a river to divert water into a canal. It raises the upstream water level of the river but does not provide any storage volume.

Drainage layer : layer of pervious material to relieve pore pressures and/or to facilitate drainage : e.g., drainage layer in an earthfill dam.

Earth dam : massive earthen embankment with sloping faces and made watertight.

Embankment : fill material (e.g. earth, rock) placed with sloping sides and with a length greater than its height.

Face : external surface which limits a structure : e.g. air face of a dam (i.e. downstream face), water face (i.e. upstream face) of a dam.

Filter : band of granular material which is graded so as to allow seepage to flow across the filter without causing the migration of the material from adjacent zones.

Flashboard : a board or a series of boards placed on or at the side of a dam to increase the depth of water. Flashboards are usually lengths of timber, concrete or steel placed on the crest of a spillway to raise the retention water level.

Flashy : term applied to rivers and streams whose discharge can rise and fall suddenly and is often unpredictable.

Flip bucket : A flip bucket or ski-jump is a concave curve at the downstream end of a spillway, to deflect the flow into an upward direction. Its purpose is to throw the water clear of the hydraulic structure and to induce the disintegration of the jet in the air.

Free-surface aeration : Natural aeration occurring at the free surface of high velocity flows is referred as free surface aeration or self-aeration.

FROUDE : William FROUDE (1810-1879) was a English naval architect and hydrodynamicist who invented the dynamometer and used it for the testing of model ships in towing tanks. He was assisted by his son Robert Edmund FROUDE who, after the death of his father, continued some of his work. In 1868, he utilised REECH's law of similarity to study the resistance of model ships.

Froude number : The Froude number is proportional to the square root of the ratio of the inertial forces over the weight of fluid. The Froude number is used generally for scaling free surface flows and open hydraulic structures It was named after William FROUDE but was first introduced by Ferdinand REECH. The number is called the Reech-Froude number in France.

Gabion : a gabion consists of rockfill material enlaced by a basket or a mesh. The word 'Gabion' ('gabion' in French, 'gabbione' in Italian) originates from the Italian 'gabbia' cage, meaning large cage ('cavea' in Latin).

Gabion dam : crib dam built up of gabions.

Gas transfer : process by which gas is transferred into or out of solution (i.e. dissolution or desorption respectively).

Ghaznavid : (or Ghaznevid) one of the Moslem dynasties (10-th to 12-th centuries) ruling South-Western Asia. Its capital city was at Ghazni (Afghanistan).

Gravity dam : dam which relies on its weight for stability. Normally the term "gravity dam" refers to masonry or concrete dam.

Hasmonean : designing the family or dynasty of the Maccabees, in Israel. The Hasmonean Kingdom was created following the uprising of the Jews in B.C. 166.

Himyarite : Important Arab tribe of antiquity dwelling in Southern Arabia (B.C. 700 to A.D. 550).

Hohokams : Native Americans in South-West America (Arizona), they build several canal systems in the Salt river valley during the period B.C. 350 to A.D. 650. They migrated to Northern Mexico around A.D. 900 where they build other irrigation systems.

Hydraulic diameter : id defined as the equivalent pipe diameter.

Hydraulic fill dam : embankment dam constructed of material which are conveyed and placed by suspension in flowing water.

Hydraulic jump : transition from a rapid (supercritical flow) to a slow flow motion. First described by LEONARDO DA VINCI, the first experimental investigation was published by Giorgio BIDONE in 1820. The present theory of the jump has been verified experimentally by BAKHMETEFF and MATZKE (1936).

Inca : South-American Indian of the Quechuan tribes of the highlands of Peru. The Inca civilisation dominated Peru between A.D. 1200 and 1532. The domination of the Incas were terminated by the Spanish conquest.

Inception of air entrainment : critical flow conditions for which free-surface air entrainment starts.

Intake : any structure in a reservoir through which water can be drawn into a waterway or pipe.

KARMAN : Theodore von KÁRMÁN (1881-1963) was a Hungarian fluid dynamicist and aerodynamicist who worked in Germany (1906 to 1929) and later in USA. He was a student of Ludwig PRANDTL in Germany. He gave his name to the vortex shedding behind a cylinder (Karman vortex street).

Left abutment : abutment on the left-hand side of an observer when looking downstream.

LEONARDO DA VINCI : Italian artist (painter and sculptor) who extended his interest to medicine, science, engineering and architecture (A.D. 1452-1519).

Lining : coating on a channel bed to provide watertightness, to prevent erosion or to reduce friction.

Lumber : timber sawed or split into boards, planks or staves.

MANNING : Robert MANNING (1816-1897) was Chief Engineer of the Office of Public Works, Ireland. In 1890, he presented two formula (MANNING 1890) : one became the 'Manning formula' but Robert MANNING did prefer to use the second formula that he gave in his paper.

Masonry dam : dam constructed mainly of stone, brick or concrete blocks jointed with mortar.

Mixing length : The mixing length theory is a turbulence theory developed by L. PRANDTL, first published in 1932. PRANDTL assumed that the mixing length is the characteristic distance travelled by a particle of fluid before its momentum is changed by the new environment.

Mochica : 1- South American civilisation (A.D. 200-1000) living in the Moche river valley, Peru along the Pacific coastline. 2- Language of the Yuncas.

Moor : 1- Native of Mauritania, a region corresponding to parts of Morocco and Algeria. 2- A Moslem of native North African races.

Morning-Glory spillway : vertical discharge shaft, more particularly the circular hole form of a drop inlet spillway. The shape of the intake is similar to a Morning-Glory flower (American native plant (Ipomocea)). It is sometimes called a Tulip intake.

Nabataean : habitant from an ancient kingdom to the East and South-East of Palestine that include the Neguev desert. The Nabataean kingdom lasted from around B.C. 312 to A.D. 106. The Nabataeans built a large number of soil-and-retention dams. Some are still in use today (SCHNITTER 1994).

Nappe flow : flow situation where the water bounces from one step to the next one as a succession of free-fall jets.

Outlet : opening through which water can be freely discharged from a reservoir to the river (e.g. bottom outlet).

Pervious zone : part of the cross-section of an embankment comprising material of high permeability.

Plunging jet : liquid jet discharging into a receiving pool of liquid.

PRANDTL : Ludwig PRANDTL (1875-1953) was a German physicist and aerodynamicist who introduced the concept of boundary layer in 1904 and developed the turbulent "mixing length" theory.

Renaissance : period of great revival of art, literature and learning in Europe in the 14-th, 15-th and 16-th centuries.

REECH : Ferdinand REECH (1805-1880) was a French naval instructor who proposed first the Reech-Froude number in 1852 for the testing of model ships and propellers.

REYNOLDS : Osborne REYNOLDS (1842-1912) was a British physicist and mathematician who expressed first the 'Reynolds number' in 1883 and later the Reynolds stress (i.e. the turbulent shear stress).

Reynolds number : dimensionless number proportional to the ratio of the inertial force over the viscous force.

Right abutment : abutment on the right-hand side of an observer when looking downstream.

RIQUET : Pierre Paul RIQUET (1604-1680) was the designer and Chief Engineer of the Canal du Midi built between 1666 and 1681. The Canal provides an inland route between the Atlantic and the Mediterranean across Southern France.

Rockfill : material composed of large rocks or stones loosely placed.

Rockfill dam : embankment dam in which more than 50% of the total volume comprise compacted or dumped pervious natural stone.

Roller : large-scale turbulent eddy.

Roller Compacted Concrete (RCC) : Roller compacted concrete is defined as a no-slump consistency concrete that is placed in horizontal lifts and compacted by vibratory rollers.

Sabaen : ancient name of the people of Yemen in Southern Arabia. Renown for the visit of the Queen of Sabah (or Sheba) to the King of Israel around B.C. 950 and for the construction of the famous Marib dam (B.C. 115 to A.D. 575). The fame of the Marib dam was such that its final destruction (in A.D. 575) was recorded in the Koran.

Seepage : interstitial movement of water that may take place through a dam, its foundation or abutments.

Side-channel spillway : A side-channel spillway consists of an open spillway (along the side of a channel) discharging into a channel running along the foot of the spillway and carrying the flow away in a direction parallel to the spillway crest (e.g. Arizona-side spillway of the Hoover dam, USA).

Skimming flow : flow regime where the water flows as a coherent stream in a direction parallel to the pseudo-bottom formed by the edges of the steps.

Slope : 1- side of a hill. 2- inclined face of a canal. 3- inclination from the horizontal.

Spillway : opening built into a dam or the side of a reservoir to release (to spill) excess flood waters.

Splitter : concrete block installed on a chute to split the flow and to increase the energy dissipation.

Staircase : another adjective for 'stepped' : e.g., a staircase cascade.

Stilling basin : A stilling basin is structure for dissipating the energy of the flow below a spillway, outlet work, chute or canal structure. In the majority of cases, a hydraulic jump is used as the energy dissipator within the stilling basin.

Stop-logs : form of sluice gate comprising a series of wooden planks, one above the other, and held at each end.

Total head : The total head is the total energy per unit mass and per gravity unit. It is expressed in metre of water.

Trashrack : screen comprising metal or reinforced concrete bars located in a waterway at an intake to precent the ingress of floating or submerged debris.

TURRIANO : Juanelo TURRIANO (1511-1585) was an Italian clockmaker, mathematician and engineer who worked for the Spanish Kings CHARLES V and later Philip II. It is reported that he checked the design of the Alicante dam for King Philip II.

Uplift : upward pressure in the pores of a material (interstitial pressure) or on the base of a structure.

VAUBAN : Sébastien VAUBAN (1633-1707), Maréchal de France, extended the feeder system of the Canal du Midi in 1686-1687.

VOC : Volatile Organic Compound.

Von Karman constant : 'universal' constant of proportionality between the Prandtl mixing length and the distance from the boundary. Experimental results indicate that $K = 0.40$.

Wadi : Arabic word for a valley which becomes a watercourse in rainy seasons.

Wake region : separation region downstream of the streamline that separates from a boundary, called a wake.

Wasteweir : a spillway. A staircase wasteweir is a stepped spillway.

Water-mill : mill (or wheel) powered by water.

Weir : low river dam used to rise the upstream water level. Measuring weir are built across a stream for the purpose of measuring the flow.

WES standard spillway shape : spillway shape developed by the US Army Corps of Engineers at the Waterways Experiment Station.

White waters : Non technical term used to design free-surface aerated flows. The refraction of the light by the entrained air bubbles gives the whitish appearance to the free-surface of the flow.

Yunca : Indian of a group of South American tribes of which the Chimus and the Chinchas are the most important. The Yunca civilisation developed a pre-Inca culture on the coast of Peru.

Acknowledgments

The author wants to thank Professor C.J. APELT (University of Queensland, Australia) for his continuous and generous support to this work. Further he wishes to acknowledge the helpful comments of Dr J.S. MONTES (University of Tasmania, Australia).

The author wishes to express his gratitude to the followings who made available some photographs of interests :

Dr R. BAKER, University of Salford, United Kingdom,

Mr Hans BANDLER, Turramurra NSW, Australia,

Mr Errol BEITZ, Queensland Water Resources, Australia,

Mr K.D. GARDINER, North West Water Engineering, United Kingdom,

Officine Maccaferri, Italy, and Maccaferri Gabions South Pacific, Australia

Mr P. ROYET, CEMAGREF, France,

Victoria Rural Water Corporation, Australia,

Mr Gordon WILLIAMS, Geolab Group, Australia,

Dr Youichi YASUDA, Nihon University, Japan.

The permission of the American Society of Civil Engineers for the reprint of some photographs is acknowledged.

The author thanks also the following people in providing some information :

Mr E.A. BATTISON, Windsor, Vermont USA,

Mr E. BEITZ, Queensland Water Resources, Australia,

Pr A. LEJEUNE, University of Liège, Belgium,

Ms QIAO G.L., Ph.D. student, The University of Queensland,

Dr P. REICHERT, EAWAG, Switzerland,

Dr M.J. TOZZI, Universidade Federal do Paraná, Brazil.

CHAPTER 1
INTRODUCTION

1.1 Presentation

Recent advances in technology have permitted the construction of large dams, reservoirs and channels. These progresses have necessitated the development of new design and construction techniques, particularly with the provision of adequate flood release facilities. Chutes and spillways are designed to spill large water discharges over a hydraulic structure (e.g. dam, weir) without major damage to the structure itself and to its environment.

Overflow spillways enable flood releases over the dam. The water flows as open channel flows or as free-falling jets, and it is necessary to dissipate the major part of its kinetic energy. Otherwise the overflowing waters could endanger the dam's toe, the surroundings and eventually the dam itself. Energy dissipation is usually achieved by : 1- a high velocity water jet taking off from a flip bucket (or from the dam crest) and impinging into a downstream plunge pool acting as a water cushion, 2- a standard stilling basin downstream of the spillway where a hydraulic jump is created to dissipate a large amount of flow energy, or 3- the construction of steps on the spillway to assist in energy dissipation.

Fig. 1-1 - Stepped spillway models

(A) Gabion spillway model (Courtesy of Mr ROYET, CEMAGREF) - Model used by PEYRAS et al. (1992) - α = 45 degrees, q_w = 0.22 m^2/s, h = 0.2 m, W = 0.8 m - Skimming flow regime

1

Fig. 1-1 - Stepped spillway models

(B) Three 1/25-scale models (Courtesy of Mr ROYET, CEMAGREF)

From the left to the right : α = 63.4, 59 and 53 degrees - h = 0.024 m

Water flowing over a rough or stepped channel can dissipate a major proportion of its energy. For a stepped chute, the steps increase significantly the rate of energy dissipation taking place along the spillway face, and eliminate or reduce greatly the need for a large energy dissipator at the toe of the spillway.

In this monograph, the first chapter describes the basic hydraulic characteristics of stepped channels and their applications. A second chapter reviews the history of stepped spillways and irrigation channels. Then, the flow characteristics and energy dissipation of stepped chute flows are detailed for nappe flow situations (chapter 3) and skimming flow regime (chapters 4 and 5). The calculations are compared with model and prototype data. Characteristics of the models and prototypes are summarised in table 1-1 and figure 1-1. The mechanisms of air-water gas transfer are discussed also (chapter 6). Later, practical examples of stepped chute design are developed (chapter 7). In the last part (chapter 8), failures and accidents are reviewed and the safety of stepped channels is discussed.

Details of nappe flow calculations are reproduced in appendix A. Appendix B lists the main physical and chemical properties of fluids. Plunging jet and skimming flow calculations are detailed in appendix C and D respectively. The monograph presents the results expressed in SI Units. Since some countries continue to use British and American units, appendix E gives their equivalents against the SI units. As human nature is never perfect ('Errare humanum est"), the reader will find a correction form in appendix F.

Each symbol is explained in the list of symbols at the beginning of the monograph. Immediately after, a glossary clarifies the main 'technical' (and non-technical) terms used in the monograph.

It must be noted that the present study describes primarily results for channels with flat horizontal steps (unless stated). ESSERY and HORNER (1978), PEYRAS et al. (1992) and FRIZELL (1992) discussed experimental results obtained with inclined and pooled steps.

Fig. 1-1 - Stepped spillway models

(C) 1/10-scale model (Courtesy of Mr ROYET, CEMAGREF) - α = 51.3 degrees, h = 0.06 m

Note the few smaller steps at the upstream end, the smooth chute on the left and the pressure tappings on the vertical faces of steps 5, 6 and 7

Table 1-1 - Characteristics of model and prototype studies

Ref.	Slope (deg.)	Scale	Step height h (m)	Nb of steps	Discharge q_w (m^2/s)	Flow regime	Remarks
(1)	(2)	(3)	(4)	(5)	(6)	(7)	(8)
Model studies							
MOORE (1943)			0.15 to 0.46	1		Nappe flow	Drop structure model (USA). W = 0.28 m.
RAND (1955)			0.2	1	0.00025 to 0.004	Nappe flow	Drop structure model (USA). W = 0.5 m.
HORNER (1969) (^{1})	22 to 40			8 to 30		Nappe flow	(UK).
ESSERY and HORNER (1978)	11 to 40		0.03 to 0.05	4 to 18		Nappe and skimming flow	CIRIA tests (UK). Include inclined steps.
ESSERY et al. (1978)	11 to 45		0.025 to 0.5	4 to 40	0.01 to 0.145	Nappe and skimming flow	CIRIA tests (UK). W = 0.15 m. H$_{dam}$ = 2 m.
STEPHENSON (1979a)	18.4 to 45		0.15	1 to 4		Nappe and skimming flow	Gabion stepped chute (South Africa). W = 0.38 m
			0.1				W = 0.1 m.
NOORI (1984)	5.7		0.004	100	0.007 to 0.09	Skimming flow	(UK). W = 0.5 m.
	11.3		0.013	70	0.025 to 0.2		
SORENSEN (1985)	52.05	1/10	0.061	11	0.006 to 0.28		Monksville dam spillway model (USA). W = 0.305 m. WES crest profile with smaller first steps.
		1/25	0.024	59	0.006 to 0.28	Nappe and skimming flow	
BAKER (1990)	21.8		0.0096 to 0.058	up to 50	up to 0.5	Skimming flow	W = 0.6 m (UK). Wedge shape blocks inclined downward : δ = - 8.3 deg.
	33.7						Flat blocks parallel to bed.
DIEZ-CASCON et al. (1991)	53.1	1/10	0.03 & 0.06	50 to 100	0.022 to 0.28	Skimming flow	H$_{dam}$ = 3.8 m (Spain). W = 0.8 m.
BAYAT (1991)	51.3	1/25	0.024		0.006 to 0.07		Godar-e-landar spillway model (Iran). W = 0.3 m.
			0.03				
			0.02				
FRIZELL and MEFFORD (1991)	26.6		0.051	47	0.077	Skimming flow	USBR Research laboratory (USA). W = 0.457 m.
STEPHENSON (1991)	54.5						Kennedy's vale model (South Africa).
PEYRAS et al. (1991,1992)	18.4, 26.6, 45	1/5	0.20	3, 4, 5	0.04 to 0.27	Nappe and skimming flow	Gabion stepped chute (France). W = 0.8 m.
BEITZ and LAWLESS (1992)	51.3 & 48.0	1/60	0.02	10	6E-4 to 0.093	Nappe and skimming flow	Burton Gorge dam spillway model (Australia). Smooth crest with smaller first steps.
FRIZELL (1992)	26.6		0.051	47	0.58	Skimming flow	USBR Research laboratory (USA). W = 0.457 m.
TOZZI (1992)	53.13	1/15	0.0083 0.0166 0.0333 0.05 0.10		0.086 to 0.201	Skimming flow	Model 1 (W = 0.7 m). WES crest profile with smaller first steps (Brazil). H$_{dam}$ = 2.1 m.

Table 1-1 - Characteristics of model and prototype studies

Ref.	Slope	Scale	Step height	Nb of steps	Discharge	Flow regime	Remarks
	(deg.)		h (m)		q_w (m²/s)		
(1)	(2)	(3)	(4)	(5)	(6)	(7)	(8)
BINDO et al. (1993)	51.34	1/21.3	0.038	31 - 43	0.01 to 0.142	Skimming flow	M'Bali spillway model (France). Creager crest profile with smaller first steps (W = 0.9 m).
		1/42.7	0.019	43	0.007 to 0.04	Skimming flow	
CHRISTODOULOU (1993)	55		0.025	15	0.02 to 0.09	Skimming flow	W = 0.5 m (Greece).
Prototype Studies GRINCHUK et al. (1977)	8.7	--	0.41	12	1.8 to 60	Skimming flow	Dneiper hydro plant, Ukraine (former USSR). Full-scale tests. W = 14.2 m.

Notes :

δ angle between the step and horizontal : $\delta > 0$ inclined upward, $\delta < 0$ inclined downward

H_{dam} model height : i.e. model crest head above downstream toe

W   channel width

$(^1)$:   as reported in CHAMANI and RAJARATNAM (1994)

1.2 Applications

For the last decades, stepped spillways have become a popular method for handling flood releases (fig. 1-2(A) and 1-2(B)). Examples of application of stepped channels are presented in table 1-2 and figure 1-2.

Recently the development of new construction materials (e.g. roller compacted concrete RCC, reinforced gabions) has increased the interest in stepped spillways. Construction of stepped spillways is compatible with slipforming and RCC placing methods (fig. 1-2(C)). Also, gabion stepped spillways are the most common type of spillways used for gabion dams (fig. 1-2(D)). The steps increase significantly the rate of energy dissipation taking place along the spillway face, and reduce the size and the cost of the downstream stilling basin.

Steps are used also to dissipate energy in stormwater channels (e.g. Canberra, Hong Kong) and for river training (fig. 1-2(E) and 1-2(F)). On Hong Kong Island, the author observed storm-waterways on steep hill slopes designed with stepped channels to assist in energy dissipation in addition of usual baffles (fig. 1-2(G) and 1-2(H)). In Norway, steps were introduced in a tunnel spillway with free-surface flow to enhance the energy dissipation. This design was selected to allow detrainment upstream of a vertical shaft and to prevent entrainment of air in the shaft (Water Power 1992).

Fig. 1-2 - Examples of stepped chutes

(A) Clywedog dam, UK (1968) (Concrete 1993)

View from downstream with waters discharging over the stepped spillways in between the dam buttresses

(B) Dartmouth dam, Australia (1977) (Courtesy of the Victoria Rural Water Corporation)

View from downstream : the dam is on the left - On the right, note the rock-lined stepped cascade and the concrete intake at the top.

Fig. 1-2 - Examples of stepped chutes

(C) RCC diversion dam near Bundaberg, Australia : Bucca weir after completion on 27 May 1987 (Courtesy of Mr BEITZ, Queensland Water Resources Commission) - Note the concrete apron downstream of the spillway toe - α = 63.4 degrees, h = 0.6 m, W = 130.8 m, H_{dam} = 11.8 m

(D) Gabion stepped weir near Bologna, Italy (1966) (Courtesy of Officine Maccaferri) - Note the concrete slab protection on the step faces - α = 45 degrees, h = 1 m, W = 11 m, H_{dam} = 5 m

Fig. 1-2 - Examples of stepped chutes

(E) River training near Macerata, Italy (Courtesy of Officine Maccaferri) - Flow from the background to the foreground - Gabion lining with steps regularly spaced

Stepped channels can be used also to increase the discharge capacity. In Russia, Soviet engineers developed the concept of overflow earth dam (GRINCHUK et al. 1977, PRAVDIVETS and BRAMLEY 1989). The spillway consists of a revetment of precast concrete blocks laid on a filter and erosion protection layer. The channel bed is very flexible and allows differential settlements : individual blocks do not need to be connected to adjacent blocks. And high discharge capacity can be achieved (table 1-2). The high degree of safety allows the use of such channels as primary spillways (PRAVDIVETS and BRAMLEY 1989). In England, a stepped intake of a Morning-glory type spillway was selected following model tests which showed that this arrangement provided a larger discharge capacity than a smooth intake (i.e. Ladybower reservoir).

In North-Eastern Australia, a number of stepped diversions weirs have been built over the past twenty years. most of these weirs comprise a downstream stepped face formed by rows of steel-

sheet piles or concrete hollow boxes with concrete capping between the rows. The stepped
profile contributes to the energy dissipation until the tailwater level rises and drowns out the
weir.

Stepped cascades are utilised also in water treatment plants. Artificial stepped cascades and
drop structures can be introduced along or beside rivers and streams to re-oxygenate waters
with low dissolved oxygen contents. Near Chicago (USA), five artificial cascades were designed
along a waterway system to help the re-oxygenation of the polluted canal (GASPAROTTO 1992).
The waterfalls were landscaped as leisure parks and combined flow aeration and aesthetics.
Aesthetical applications of stepped cascades include stepped fountains in cities (e.g in Brisbane,
Hong Kong, Taipei, Tokyo) (fig. 1-2(I) and 1-2(J)).

Fig. 1-2 - Examples of stepped chutes

(F) Training of a torrent near La Paz, Bolivia (1986) (Courtesy of Officine Maccaferri) - Gabion
steps with lumber protection at the edges of the steps

Table 1-2 Typical examples of stepped cascades and chutes

Name	Ref.	Slope α (deg.)	Dam height H_{dam} (m)	Max. disch. q_w (m²/s)	Step height h (m)	Nb of steps	Type of steps	Remarks
(1)	(2)	(3)	(4)	(5)	(6)	(7)	(8)	(9)
Masonry dam spillway								
Gilboa dam, USA 1926	[GM]		49	7.8	6.1	8	Steps inclined downwards : δ = -2.9 to -5.7 deg.	W = 403.6 m. 9 steps.
Concrete dam spillway								
Clywedog dam, UK 1968	[E2]	60	72	2.8	0.76		Precast concrete beams.	Buttress dam. Precast beam spillway. W = 182.9 m.
De Mist Kraal weir, South Africa 1986	[HD]	59	30	29	1	19	Horizontal steps.	RCC dam. W = 195 m.
Zaaihoek dam, South Africa 1986	[HD]	58.2	45	15.6	1	40	Horizontal steps.	RCC dam. W = 160 m.
Monksville, USA 1987	[SO]	52	36.6	9.3	0.61		Horizontal steps.	RCC dam. W = 61 m.
Olivettes dam, France 1987	[BO, GO]	53.1	36	6.6	0.6	47	Horizontal steps.	RCC dam. W = 40 m.
Upper Stillwater, USA 1987	[HO]	72 & 59	61	11.6	0.61		Horizontal steps.	Overflow RCC dam. W = 183 m.
M'Bali dam, RCA 1990	[BI]	51.3	24.5	16	0.8	36	Horizontal steps.	RCC dam. W = 60 m.
New Victoria dam, Australia 1993	[WA]	72 & 51.3	52	5.4	0.6	85	Horizontal steps.	Overflow RCC dam. W = 130 m.
Petit-Saut dam, Guyana 1994	[DU, GO]	51.3	37	4	0.6		Horizontal steps.	RCC dam.
Earth overflow dam								
Brushes Clough dam, UK 1859/1991	[BG]	18.43	26	Q_w = 3.66 m³/s	0.19		Wedge concrete blocks inclined downward : δ = - 5.6 deg.	New overflow (1991). Trapezoidal section (2-m wide, 1V:2H sideslope).
Dneiper hydro plant, Ukraine (former USSR) 1976	[PR]	8.75		60	0.405	12	Concrete block system. Horizontal steps.	Full scale tests. V = 23 m/s. W = 14.2 m.
Sosnovski dam, Russia (former USSR) 1978	[PR]		11	3.3			Stepped block system.	W = 12 m.
Lukhovitsky dams, Russia (former USSR) 1978, 1980 and 1981	[KR]		11, 11.5, 7.5	3, 3.3, 2.9			Stepped block system.	3 earthen dams. W = 12, 12, 7.5 m.
Transbaikal region dam, Russia (former USSR) 1986	[MIL]	14	9.4	20			Stepped block system.	Reservoir capacity : 1.5 Mm³. W = 110 m.
Gabion dam								
Rietspruit outfall, South Africa	[ST]			13		2 to 4	Gabion steps.	3 successive stepped weirs. W = 50 m.
San Paolo weirs Brazil 1986	[AG]	63	9, 11 and 8		1	9, 11 and 8	Horizontal gabion steps.	Solid transport and flood flow controls. 3 successive weirs.

Table 1-2 Typical examples of stepped cascades and chutes

Name	Ref.	Slope α (deg.)	Dam height H_{dam} (m)	Max. disch. q_w (m^2/s)	Step height h (m)	Nb of steps	Type of steps	Remarks
(1)	(2)	(3)	(4)	(5)	(6)	(7)	(8)	(9)
Tunnel spillway Stoyord tunnel system, Norway 1993	[WP]	11.3		$Q_w = 120$ m^3/s		4	22	Pooled steps with 1-m high wall.	Flow de-aeration U/S of a vertical shaft. Length : 420 m.
Unlined rock spillway Bellfield dam, Australia 1966	[MIC, T]		54.9	20.9	12.2	3	Unlined cascade. Flat steps.	Earth-cored rockfill-flanked dam. W = 23 m.
Dartmouth dam, Australia 1977	[T]		180	$Q_w = 2700$ m^3/s	5	13	Unlined cascade in granite. Flat steps.	Earth and rockfill dam. Channel width varies from 91.4 m at crest up to 350 m.
La Grande 2 dam, Canada 1982	[PO]	30	134	$Q_w = 16140$ m^3/s	17.8	1	Smooth profiled step. Pool depth = 8.5 m.	W = 122 m. 1st step : concrete lined.
					9.1 to 12.2	11	Unlined horizontal steps.	Unlined excavations. 2nd to 12th steps.
Tehri dam spillway, India	[VI]	15	211.9	$Q_w = 11000$ m^3/s	50 to 58	4	Smooth pooled steps. Pool depths from 14 to 18 m.	Smooth profiled cascade system. 4 smooth drops. W = 80 to 95 m.
Paunglaung dam spillway, Burma	[LY]	12	120	$Q_w = 10000$ m^3/s	25 to 35	4	Pooled steps with 10-m high walls.	Unlined rock.
Debris dam Harvey Creek, Canada	[HU]					16	Flat steps.	Stepped debris spillway and double culvert.
Yuba River debris barrier, USA 1904	[E4]	15.4	6.8	9.8	1.8 to 2.6	3	Smooth profiled steps. Concrete facing.	Hydraulic fill dam anchored by rockfill, piling and concrete facing. W = 380 m.
Timber and crib dam Arizona canal dam, USA 1887	[E3]		10	33		3	Horizontal steps.	Timber cribs filled with loose rock and gravel.
Canyon Ferry dam, USA 1898	[WE]		8.8		2.3 to 3.05	3	Horizontal steps.	Timber cribs filled with stones. W = 148 m.
Feather river diversion weir, USA 1912	[ET]	26.6	7.9	73	2.4	4	Horizontal steps.	Crib weir. W = 85.3 m.
Diversion weirs Bucca weir, Australia 1987	[QW]	63.4	11.8	55.4	1.6	1	Flat crest (4.5-m long) and first step.	Overflow diversion dam (W = 130.8 m). RCC structure.
					0.6	17	Inclined upwards steps : $\delta = +2.9$ deg.	

Name	Ref.	Slope α (deg.)	Dam height H_{dam} (m)	Max. disch. q_w (m^2/s)	Step height h (m)	Nb of steps	Type of steps	Remarks
(1)	(2)	(3)	(4)	(5)	(6)	(7)	(8)	(9)
Diversion weirs								
Maranoa river weir, Australia 1984	[MIK]	14.5	4.7	~ 16	1.5	4	Horizontal steps with nib wall dissipators.	Interlinked concrete boxes filled with rockfill covered by concrete slabs. W = 90 m.
Fountains								
Peak tramway fountain, Hong Kong		20	1.8		0.15	12	Horizontal steps.	Nappe flow regime.
Taipei world trade centre fountain, Taiwan							Horizontal steps.	Nappe flow regime
Re-aeration cascade								
Calumet waterway system, USA 1991	[GA]				1.52 & 0.91	3 & 4	Waterfalls.	5 artificial waterfall cascades.
Tulip intake of Morning Glory type spillway								
Ladybower reservoir, UK 1939	[E1]				0.46	16	Masonry steps.	Improve discharge efficiency of the overflow bellmouth.
Storm waterway								
Hatton Rd, Hong Kong		50	4		0.4	15	Steps and baffle blocks.	W = 4 m.
		10	10		0.2	35		

Note : [AG] AGOSTINI et al. (1987); [BI] BINDO et al. (1993); [BG] BAKER and GARDINER (1994); [BO] BOUYGE et al. (1988); [DU] DUSSART et al. (1993); [E1] The Engineer (1939); [E2] Engineering (1966); [E3] Engineering News (1905a); [E4] Engineering News (1905b); [ET] ETCHEVERRY (1916); [GA] GASPAROTTO (1991); [GM] GAUSMANN and MADDEN (1923); [HD] HOLLINGWORTH and DRUYTS (1986); [HO] HOUSTON and RICHARDSON (1988); [HU] HUNGR et al. (1987); [KR] KREST'YANINOV and PRAVDIVETS (1986); [LY] LYSNE (1992); [MIC] MICHELS (1966); [MIK] MICKELSON and MOORWOOD (1984); [MIL] MILLER et al. (1987); [PO] POST et al. (1987); [PR] PRAVDIVETS and BRAMLEY (1989); [QW] Queensland Water Resources (1988); [SO] SORENSEN (1985); [ST] STEPHENSON (1979a); [T] THOMAS (1976); [VI] VITTAL and POREY (1987); [WA] WARK et al. (1991); [WP] Water Power (1992); [WE] WEGMANN (1911).

Another application is the debris[1] torrent check dams. In mountain areas, debris[1] torrents may

---

[1]Debris comprise mainly large boulders, rock fragments, gravel-sized to clay-sized material, tree and wood material that accumulate in mountain creeks (VANDINE 1985).

have catastrophic (and dramatic) impacts. "Check dams" (or debris dams) can be used to reduce the impact of debris torrents (VANDINE 1985). Their construction near the headwaters of the creeks may prevent the "initiation" of the debris flow. Checks dams are usually constructed as a succession of drop structures (h = 0.5 to 5 m) to reduce the steep gradient of the creek, to enhance the energy dissipation and to prevent the onset of debris flow (fig. 1-2(K)). Some debris dams are designed also with culverts and stepped overflow spillway (e.g. Harvey Creek). Check dams are common in Europe (Austria, France, Italy, Poland, Switzerland), North America and Far East Asia (Japan, Taiwan) (e.g. MAMAK 1964, VANDINE 1985, Soil and Water 1992).

Fig. 1-2 - Examples of stepped chutes

(G) Stepped storm water way, Hong Kong in 1993 - Intake of the stepped storm waterway

α = 50 degrees, h = 0.4 m, W = 4 m

Fig. 1-2 - Examples of stepped chutes

(H) Stepped storm water way, Hong Kong in 1993 - α = 10 degrees, W = 4 m

Note the baffle blocks at the end of the channel

(I) Stepped fountain, Peak tram cascade, Hong Kong in 1993

Fig. 1-2 - Examples of stepped chutes

(J) Stepped fountain, Mitsui Shinjiku Building, Tokyo in 1993

(K) Series of six check dams to control the solid transport in the Shaar Descent, Saudi Arabia
(1982) (Courtesy of Officine Maccaferri)

Note the concrete capping to protect the crest of the drop structures

1.3 Flow regimes

A stepped chute consists of an open channel with a series of drops in the invert. The flow over stepped spillways can be divided into two regimes : nappe flow and skimming flow (fig. 1-3).

In a *nappe flow regime* (fig. 1-3(A)), the total fall is divided into a number of smaller free-falls. The water proceeds in a series of plunges from one step to another (fig. 1-4). The flow from each step hits the step below as a falling jet followed by a hydraulic jump in most cases. The energy dissipation occurs by jet breakup in air, by jet mixing on the step, and with the formation of a fully developed or partial hydraulic jump on the step (RAJARATNAM 1990). Over small dams, a large rate of energy dissipation can take place (ELLIS 1989, PEYRAS et al. 1992). For a nappe flow regime, the steps need to be relatively large (paragraph 3.5). This situation is not often practical but may apply to relatively flat spillways, streams and stepped channels.

In a *skimming flow regime*, the water flows down the stepped face as a coherent stream, skimming over the steps and cushioned by the recirculating fluid trapped between them (fig. 1-3(B)). The external edges of the steps form a pseudo-bottom over which the flow passes. Beneath this, recirculating vortices develop and are maintained through the transmission of shear stress from the water flowing past the edge of the steps. In the early spillway steps, the flow is smooth and no air entrainment occurs. After a few steps the flow is characterised by a strong air entrainment and by vortices at the step toes (fig. 1-5). The energy dissipation in the flow appears to be enhanced by the momentum transfer to the recirculating fluid (RAJARATNAM 1990). It will be shown that, on large dams, skimming flow situations dissipate more energy than nappe flow situations (Chapter 5).

The transition between nappe flow and skimming flow regime is discussed in chapter 4 (paragraph 4.2).

1.4 Air entrainment on stepped chutes

1.4.1 Introduction

The flow conditions above a stepped channel are characterised by a high level of turbulence and large quantities of air are entrained. Air entrainment is caused by turbulence velocities acting next to the air-water free surface. Through this interface, air is continuously trapped and released. Air entrainment occurs when the turbulent kinetic energy is large enough to overcome both surface tension and gravity effects. The turbulent velocity normal to the free surface v' must overcome the surface tension pressure (ERVINE and FALVEY 1987), and be greater than the bubble rise velocity component for the bubbles to be carried away. These conditions are :

$$v' > \sqrt{\frac{8 * \sigma}{\rho_w * d_{ab}}} \qquad (1\text{-}1)$$

and $\quad v' > u_r * \cos\alpha \qquad (1\text{-}2)$

Fig. 1-3 - Flow regimes above a stepped chute

(A) Nappe flow regime with fully-developed hydraulic jump

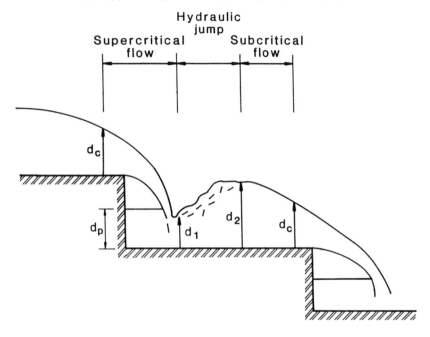

(B) Skimming flow regime with stable cavity recirculation

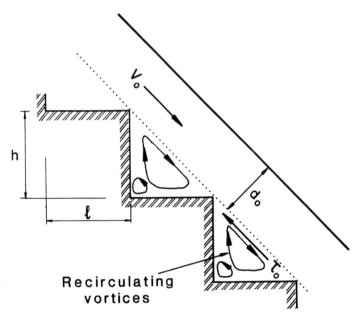

Fig. 1-4 - Example of nappe flow regime - Gabion stepped model used by PEYRAS et al. (1992)

(Courtesy of Mr ROYET, CEMAGREF) - Note the steps inclined upwards

$\alpha = 18.4$ degrees, $q_w = 0.054$ m^2/s, h = 0.2 m, W = 0.8 m, H$_{dam}$ = 4 m

Fig. 1-5 - Examples of skimming flow regime

(A) Nihon university model (Courtesy of Dr YASUDA, Nihon University) - Note the air bubbles

trapped in the recirculating cavities formed by the steps (stations 7 and 6)

$\alpha = 45$ degrees, $q_w = 0.009$ m^2/s, W = 0.4 m

Fig. 1-5 - Examples of skimming flow regime

(B) Skimming flow over three 1/25 scale models (Courtesy of Mr ROYET, CEMAGREF)

From the left to the right : α = 63.4, 59 and 53 degrees - h = 0.024 m

where σ is the surface tension, ρ_w is the water density, d_{ab} is the air bubble diameter, u_r is the bubble rise velocity and α is the channel slope.

Air entrainment occurs when the turbulent velocity v' satisfies both equations (1-1) and (1-2). Figure 1-6 shows both equations (1-1) and (1-2) for bubble sizes in the range 0.1-1000 mm and slopes from 0 to 75 degrees. The rise velocity of individual bubbles in still water was computed as by the method of COMOLET (1979b). Figure 1-6 suggests that air entrainment at the free surface occurs for turbulent velocities v' greater than 0.1 to 0.3 m/s and air bubbles in the range 8-40 mm are the most likely to be entrained. For steep slopes the action of the buoyancy force is reduced and larger bubbles are expected to be carried away.

On stepped spillways the level of turbulence is high. Both conditions (eq. (1-1) and (1-2)) are satisfied and large quantities of air are entrained along the chute.

Fig. 1-6 - Critical turbulent velocity v' for free-surface aeration (after CHANSON 1992a)

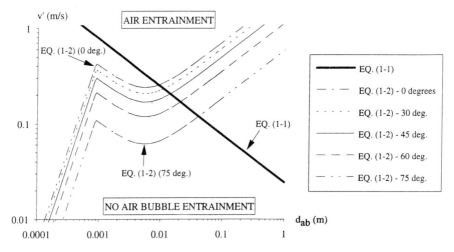

1.4.2 Air entrainment on stepped chutes

Considering a stepped chute with a nappe flow regime, air is entrained at each step by a plunging jet mechanism at the intersection of the overfalling jet and the receiving waters, and at the toe of the hydraulic jump (fig. 1-7(A)). With deep pooled steps, most of the air is entrained by the plunging jet mechanism. Extensive studies of plunging jet entrainment were performed (e.g. VAN DE SANDE and SMITH 1973,1976; ERVINE et al. 1975,1980,1987). Recent reviews include WOOD (1991), CHANSON and CUMMINGS (1992) and BIN (1993). For flat steps with shallow waters, most of the air is entrained at the toe of the hydraulic jumps. The air entrainment characteristics of hydraulic jumps were analysed by a number of researchers (e.g. RAJARATNAM 1967, RESCH et al. 1974).

On a stepped spillway with skimming flow, the entraining region follows a region where the flow over the spillway is smooth and glassy. Next to the boundary however turbulence is generated and the boundary layer grows until the outer edge of the boundary layer reaches the surface (fig. 1-7(B)). When the outer edge of the boundary layer reaches the free surface, the turbulence can initiate natural free surface aeration (CHANSON 1993c). The location of the start of air entrainment is called the point of inception. Downstream of the inception point of free-surface aeration, the flow becomes rapidly aerated and the free-surface appears white as shown by photographs (KREST'YANINOV and PRAVDIVETS 1986, DIEZ-CASCON et al. 1991, fig. 4-2 and 4-7). Far downstream the flow will become uniform and for a given discharge any measure of flow depth, air concentration and velocity distributions will not vary along the chute. This region is defined as the uniform equilibrium flow region.

Fig. 1-7 - Air entrainment on stepped spillways

(A) Flow aeration in nappe flow regime with fully-developed hydraulic jump

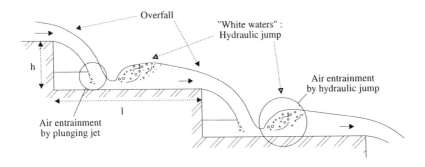

(B) Flow aeration in skimming flow regime

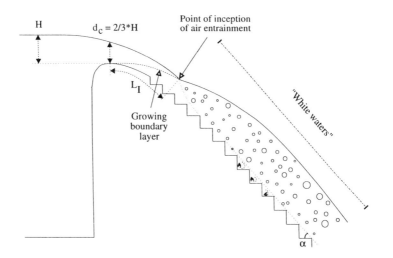

1.4.3 Effects of air entrainment

In supercritical flows above chutes and spillways, the amount of entrained air is an important design parameter. Self-aeration[2] was initially studied because of the effects of entrained air on the volume of the flow. Air entrainment increases the bulk of the flow which is a design

---

[2]Natural aeration occurring at the free surface of high velocity flows is referred as free surface aeration or self-aeration.

parameter that determines the height of spillway sidewalls (FALVEY 1980).

Also the presence of air within the boundary layer reduces the shear stress between the flow layers and hence the shear force. The resulting increase of flow momentum must be taken into account when designing a ski jump and a stilling basin downstream of a spillway, a stepped spillway or a rockfill dam with possible overtopping.

Further the presence of air within high-velocity flows may prevent or reduce the damage caused by cavitation (MAY 1987). Cavitation erosion may occur on stepped spillways. But the risks of cavitation damage are reduced by the flow aeration. PETERKA (1953) and RUSSELL and SHEEHAN (1974) showed that 4 to 8% of air concentration next to the spillway bottom may prevent cavitation damage on concrete surfaces. On stepped spillways, the high rate of energy dissipation reduces the flow momentum in comparison with a smooth chute. The reduction of flow velocity and the resulting increase of flow depth reduce also the risks of cavitation as the cavitation index increases.

Recently air entrainment on spillways and chutes has been recognised also for its contribution to the air-water transfer of atmospheric gases such as oxygen and nitrogen (WILHELMS and GULLIVER 1989, CHANSON 1993d). This process must be taken into account for the re-oxygenation of polluted streams and rivers, but also to explain the high fish mortality downstream of large hydraulic structures.

For the designer the flow bulking is estimated from the total quantity of air entrained while the prevention of cavitation damage requires the knowledge of the air concentration in the layers close to the spillway bottom. The reduction of the drag observed with air entrainment will reduce the energy dissipation above the spillway and hence its efficiency. It must be emphasised that the reduction of the friction factor observed with increasing mean air concentration is not completely understood. The presence of bubbles across the flow and the bubble size distribution is expected to affect the turbulence and the turbulent mixing mechanisms (CHANSON 1992a).

The effects of air entrainment on stepped spillway flows are currently under active study in a number of centres around the world (DIEZ-CASCON et al. 1991, FRIZELL and MEFFORD 1991, CHANSON 1993c).

CHAPTER 2
HISTORY OF STEPPED CHANNELS AND WEIRS

2.1 Introduction

Some studies (e.g. SORENSEN 1985) suggested that the design of stepped channels for energy dissipation purposes was a new technique, developed with the recent introduction of new construction materials (e.g. RCC, strengthened gabions). Indeed the construction of stepped spillways is facilitated by the slipforming and placing methods of roller compacted concrete and with the construction techniques of gabion weirs.

But the author will demonstrate that the design of stepped channels has been known since the Antiquity. Stepped chutes were designed to contribute to the stability of a structure (e.g. overflow weir) and to dissipate flow energy. In a first part, several examples of ancient stepped spillways and irrigation channels with stepped profiles will be presented. Then it will be shown that the technique of stepped channels was developed independently by several ancient civilisations. Later, the author will review the hydraulic characteristics of stepped spillways through History.

2.2 History of stepped chutes

The world's oldest stepped spillways are presumably those of the two Khosr River dams (or Ajilah dams), in Iraq. The dams were built around B.C. 694 by the Assyrian King Sennacherib. They were designed to supply water to the Assyrian capital city Nineveh, near the actual Mosul, via two canals. Both dams feature a stepped downstream face and were intended to discharge the river over their crests. Remains of these dams are still in existence (SMITH 1971, SCHNITTER 1994).

Later, the Romans built stepped overflow dams in their empire : remains are found still in Syria, Libya and Tunisia (table 2-1, fig. 2-1(A) and 2-2(A)). After the fall of the Roman empire, Moslem civil engineers gained experience from the Nabataeans, the Romans and the Sabaens. They built dams with stepped overflow in Iraq, Saudi Arabia and Spain (e.g. Adheim dam, Mestella weir). Some structures included also bottom outlets.

After the reconquest of Spain, Spanish engineers continued to use the Roman and Moslem structures. They designed also new weirs and dams with overflow stepped spillways (e.g. Almansa dam, Alicante dam, Barrarueco de Abajo dam). In 1791, they built the largest dam with an overflow stepped spillway, the Puentes dam. But the dam was washed out after a foundation failure in 1802. Before 1850, the dam expertise of Spanish engineers was most exceptional all over the world. It was exported to the "New Indies" after the conquest of America. In Central Mexico, several stepped overflow dams were built by the Spanish during the 18-th and 19-th centuries (fig. 2-2(B), 2-2(C)) and some were still in use at the beginning of the 20-th century (fig. 2-3).

By the middle of the 17-th century, French engineers knew about the Spanish experience. In the South-West of France, the feeder system of the Canal du Midi[1], designed by RIQUET and extended by VAUBAN, included several stepped channels and cascades (ROLT 1973). The stepped chutes were designed to dissipate the flow energy and to prevent scouring.

In England, several dams were built near furnaces and water mills. Some included stepped weirs and spillways. It is believed that English engineers gained some experience from the Romans who built aqueducts and dams during their occupation.

It is worth mentioning some relatively ancient timber and crib dams with stepped overflows. The North-East part of America benefited from the experience of Northern European settlers and timber dams were reported as early as A.D. 1600. During the period 1800-1920, timber dams were popular in America, Australia and New Zealand. Most timber dams were 3 to 6-m high, but some much bigger ones were built successfully to a height of 30 m. Timber overflow stepped weirs were able to sustain large flood discharges without major damage : e.g. the diversion dam on the Feather river (table 2-3).

2.3 Ancient irrigation canal systems

Drop structures and stepped profiles were used in some early irrigation systems (table 2-2, fig. 2-1(B)). Most cases suggest that the hydraulic expertise was developed locally through evolution rather than technology transfer. In Yemen and Israel, the inhabitants benefited from some local 'savoir-faire' : i.e., the Sabaen and Nabataean expertise respectively. In Arizona, Native Americans used stepped profiles in their canals before the arrival of the first Europeans (HODGE 1893). In Peru, the Indians civilisations used stepped channels and drop structures prior to the Spanish conquest.

Fig. 2-1 - Historical development of stepped channels in the World - Legend

Symbol	Dams	Symbol	Irrigation systems
O	Khosr river dams	●	Wadi Beihan valley
□	Roman dams	◆	Na'aran channel
Δ	Moslem dams	▼	Hohokam works
∇	Spanish dams	■	Peruvian irrigation systems

[1]The Canal du Midi was built between 1666 and 1681 to provide an inland route between the Atlantic and the Mediterranean across Southern France. The feeder system was later extended in 1686-1687.

Fig. 2-1 - Historical development of stepped channels in the World

(A) Ancient stepped spillways

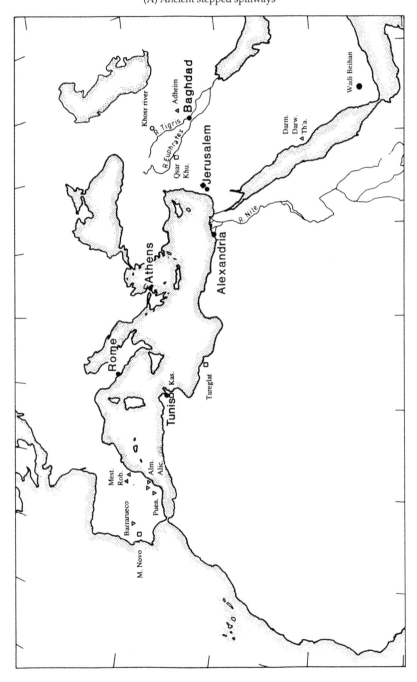

Fig. 2-1 - Historical development of stepped channels in the World

(B) Ancient irrigation systems with stepped profiles

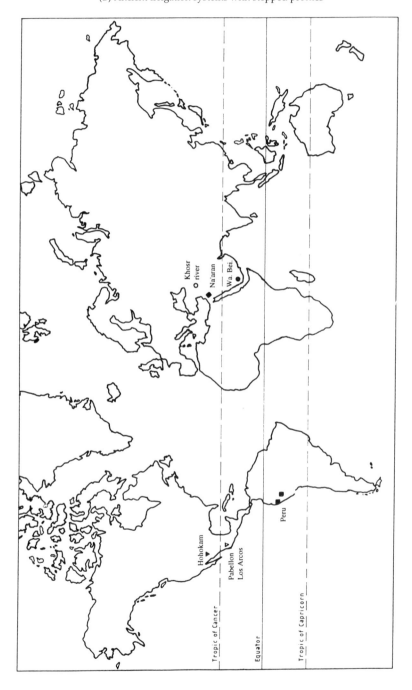

Fig. 2-2 - Ancient dams with stepped overflow spillway

(A) Sketch of the Kasserine dam, Tunisia (AD 100) : details of the spillway

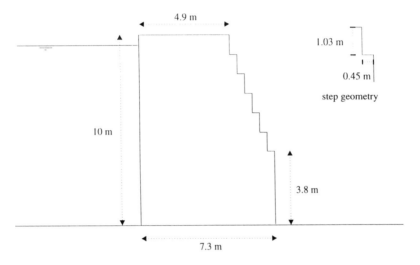

(B) Spanish dam in Mexico : Pabellon dam 1730? (after HINDS 1932)

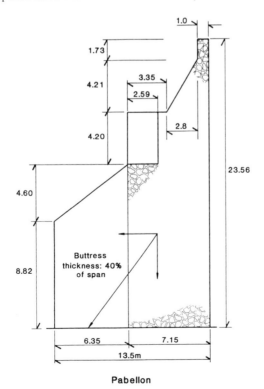

Pabellon

Fig. 2-2 - Ancient dams with stepped overflow spillway

(C) Spanish dam in Mexico : Presa de Los Arcos 1780? (after HINDS 1932)

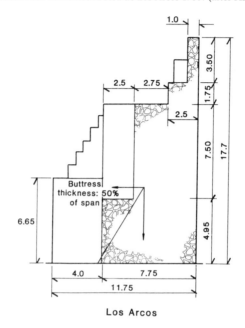

Los Arcos

(D) Quinson dam (France, 1870) (after WEGMANN 1911)

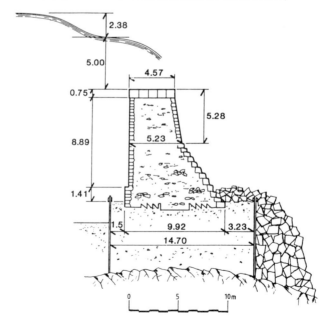

Fig. 2-2 - Ancient dams with stepped overflow spillway

(E) Tytam dam (Hong Kong 1887)

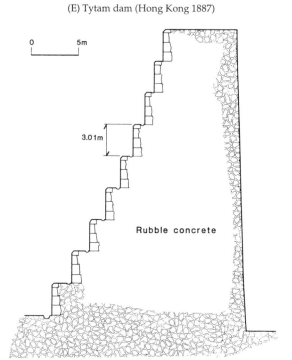

(F) Castlewood dam, Colorado (USA, 1890) - Spillway cross-section (after SCHUYLER 1909)

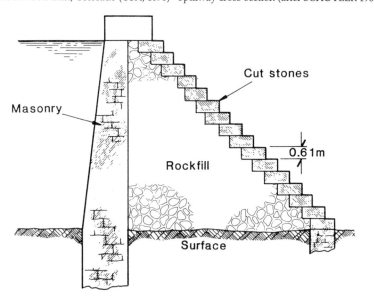

Fig. 2-2 - Ancient dams with stepped overflow spillway

(G) Pedlar River dam, Virginia, 1905 (USA) - Spillway cross-section with details of the granite
step facing (after SCHUYLER 1909)

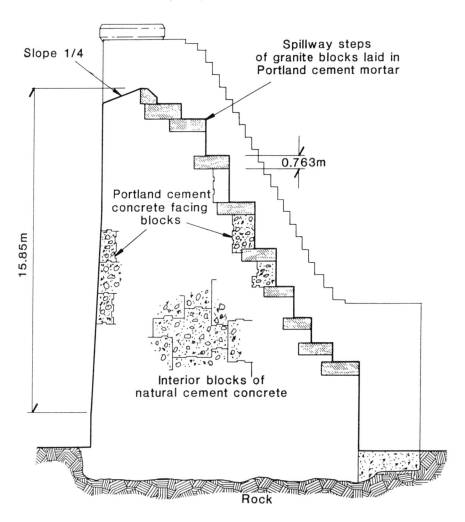

Fig. 2-2 - Ancient dams with stepped overflow spillway

(H) Croton Falls dam (USA, 1911) - Spillway cross-section (after WEGMANN 1911)

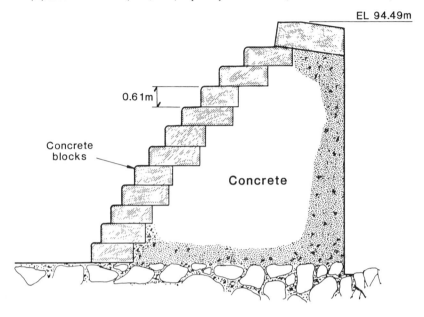

Fig. 2-3 - Spanish dams in Mexico with stepped overflow spillway

(A) Pabellon dam (Mexico, 1730?) (HINDS 1953, with permission of the American Society of Civil Engineers) - View from the left bank - Note the old water mill in the foreground and the overflow spillway on the right half of the dam

Fig. 2-3 - Spanish dams in Mexico with stepped overflow spillway

(B) Pabellon dam (Mexico, 1730?) (HINDS 1953, with permission of the American Society of Civil Engineers) - Detail of the buttresses

(C) Pabellon dam (Mexico, 1730?) (GOMEZ-PEREZ 1942, with permission of the American Society of Civil Engineers) - Waters discharging over the spillway with the old water mill on the right of the picture

Fig. 2-3 - Spanish dams in Mexico with stepped overflow spillway
(D) Presa de Los Arcos dam (Mexico, 1780?) (HINDS 1953, with permission of the American
Society of Civil Engineers) - View from the right bank.

Fig. 2-4 - Ascutney Mill dam, Windsor (USA, 1834-35) : view of a spill from the right bank
(Courtesy of the American Society of Civil Engineers) - Note the smooth crest followed by the
granite steps

Table 2-1 - Historical stepped spillways (before A.D. 1850)

Name	Year	Ref.	Dam height (m)	Slope α (deg.)	Construction	Comments
(1)	(2)	(3)	(4)	(5)	(6)	(7)
River Khosr dams, Iraq	B.C. 694	[SM, FO, TH]	2.9		Masonry of limestone, sandstone mortared together.	Built by the Assyrian King Sennacherib to supply water to his capital city Nineveh. Discharge over the dam crest. Lower dam. Called Ajilah dams [SC2].
			> 1.4	30		Upper dam. 5 steps.
Kasserine dam, Tunisia	A.D. 100?	[SM]	10	57	Cut and fitted masonry blocks with mortared joints used to face a rubble and earth core.	Roman dam 220 km SW of Tunis, Tunisia. 6 steps followed by an overfall. Discharge over the dam crest. W = 150 m.
Qasr Khubbaz, Syria	A.D. 100/200	[SM, ST]	6.1		Masonry dam with limestone slabs.	Roman dam on the Euphrates river. Reservoir capacity : 9,000 m^3 of water.
Tareglat dam, Libya	A.D. 200/300	[VI]	> 2		Masonry gravity dam reinforced by buttressing.	Roman dam near Al Khums. Soil-and-water retention dam. Step height : h = 0.6 m.
Monte Novo dam, Portugal	A.D. 300?	[QU, SC1]	5.7		Curved gravity wall reinforced by two downstream buttresses. Masonry construction.	Roman or post-Roman dam 17 km Est of Evora. Water supply to a watermill. 6 steps.
Dama dam, Saudi Arabia	A.D. 700?	[SC2]	7.5 (?)	78	Both outer walls made of dry masonry with earth or rubble core in between.	Moslem dam near Mecca. 5 steps (h ~ 1.45 m).
Darwaish dam, Saudi Arabia	A.D. 700?	[SC2]	10	71	Both outer walls made of dry masonry with earth or rubble core in between.	Moslem dam near Mecca. 5 steps (h ~ 1.6 m).
Tha'laba dam, Saudi Arabia	A.D. 700?	[SC2]	9	80	Both outer walls made of dry masonry with earth or rubble core in between.	Moslem dam near Mecca. 5 steps (h ~ 1.8 m).
Robella dam, Spain	A.D. 900?	[SC2]	3.5		Rubble masonry. Shaped crest followed by 5 steps.	Moslem stepped weir. 5 steps (h = 0.37 to 0.7 m).
Mestella dam, Spain	A.D. 960	[SM]	2.1	27	Rubble masonry and mortar core faced with large masonry blocks and mortared joints.	Stepped weir built by the Moslems. Maximum discharge : around 4,000 m^3/s (?). W = 73 m. 5 steps (h = 0.35 and 1 m).
Khan dam, Uzbekistan	1000?	[SC2]	15.2	81	Gravity dam. Granite ashlar masonry.	Moslem dam (Ghaznavid dynasty) 100 km North of Samarkand. 2.3-m wide crest followed by 7 steps (h = 1.5 to 3.3 m).
Adheim dam, Iraq	1300?	[SM]	15.2	51	Gravity dam. Cut masonry blocks connected with lead dowels poured into grooves.	Built by the Moslems during the Sassanian period. 7.5-m wide crest followed by steps.
Almansa dam, Spain	1384?	[SM, WE]	15	40	Curved gravity dam. Rubble masonry with a facing of large masonry blocks.	Discharge over the dam crest. Broad crest followed by 14 steps and an overfall.

Table 2-1 - Historical stepped spillways (before A.D. 1850)

Name	Year	Ref.	Dam height (m)	Slope α (deg.)	Construction	Comments
(1)	(2)	(3)	(4)	(5)	(6)	(7)
Ashburnham Furnace dam, UK	1563?	[BI]	7		Stones with brick and brickwork repairs.	Earth dam with gated masonry spillway. W = 3.7 m. 3 steps (h ~ 1.5 m).
Alicante dam, Spain	1594	[SM]	41	79	Curved gravity dam. Rubble masonry set in mortar, faced with masonry blocks.	Overflow spillway (W = 80 m). 7 steps : h = 2.7 to 5 m, l = 0.6 to 0.9 m.
St Ferréol dam, France	1671	[RO]	32		Waterfalls, cascades, cataracts.	Earth dam with masonry spill weir followed by stepped cascades. Water supply for the Canal du Midi.
Barrarueco de Abajo, Spain	17??	[GD]	6.3	~ 25	Buttress dam. Masonry construction.	Overflow spillway. 5 steps inclined downward.
Pabellon dam, Mexico (fig. 2-3 (A))	1730?	[GP, HI, SM]	24		Buttress dam. Rubble masonry set in mortar.	Spanish construction. Discharge over the crest. 3 steps.
Presa de los Arcos, Mexico (fig. 2-3(D))	1780?	[HI, SM]	18		Buttress dam. Rubble masonry set in mortar.	Spanish dam across the Rio Morcinique. Overflow spillway. 4 steps.
Puentes dam, Spain	1791	[SM, WE]	50	51	Gravity dam. Rubble masonry core set in mortar and faced with large cut stones. Dam failure in 1802.	Dam across the Rio Guadalentin. Discharge over the crest. 4 steps followed by an uniform slope. Step height : h = 4.175 m.
Penarth weir, UK	1818	[BI]	2.4		Masonry crest.	Stepped weir on the river Severn. 2 steps (h ~ 1.2 m). W = 42 m.
Ascutney Mill dam, USA (fig. 2-4)	1834-35	[BA, SC3]	12.8		Arched gravity dam made of cut granite with rubble filled core. Downstream stepped buttressing wall.	Overflow stepped spillway (W = 30.48 m) across the Connecticut river. Smooth concrete crest followed by stone stepped profile. Water supply to watermills and later hydropower (1898).

Note : [BA] BATTISON (1975); [BI] BINNIE (1987); [FO] FORBES (1955); [GD] GARCIA-DIEGO (1977); [GP] GOMEZ-PEREZ (1942); [HI] HINDS (1932,1953); [QU] QUINTELA et al. (1987); [RO] ROLT (1973); [SC1] SCHNITTER (1991); [SC2] SCHNITTER (1994); [SC3] SCHODEK (1987); [SM] SMITH (1971); [ST] STEIN (1940); [TH] THOMPSON and HUTCHINSON (1929); [VI] VITA-FINZI (1961); [WE] WEGMANN (1911).

One of the most ingenious systems was the Quishuarpata canal (Peru). The canal included two steep chutes over 100-m long each. Each chute was designed with small steps (h = 1 to 3 cm) along the chute course and large steps (i.e. drop structures) near the downstream end. No stilling ponds were used. These techniques of the canal highlight the hydraulic expertise of ancient Peruvian engineers (pre-Inca and Inca) : they designed canals able to sustain supercritical flows with steep slopes, drop structures and hydraulic jumps at the downstream end of the canals.

Table 2-2 - Ancient irrigation systems with stepped profile (before A.D. 1850)

Name	Year	Ref.	Disch q_w (m^2/s)	Slope α (deg.)	Construction and hydraulic design	Comments
(1)	(2)	(3)	(4)	(5)	(6)	(7)
Wadi Beihan valley, Yemen Hajar Bin Humeid	B.C. 1000 to A.D. 200	[LE]	2.2 to 4.4	0.03 to 0.11	Paved channel.	Himyarit irrigation system in Qataban. Main canal. Maximum water depth : 1.5 m.
			0.4 to 0.7	35 to 60	Stepped intake and drop structures at downstream end.	Secondary canals. W = 0.5 to 5 m. Maximum flow depth < 0.5 m. h = 0.15 to 0.3 m.
Na'aran channel, Jericho, Israel	B.C. 103 to 76	[NE]			Field stones bonded by lime mortar. Energy dissipation by drop structures.	Part of the Wadi Qelt system built by the Hasmonean King Alexander Janneus. W = 0.6 m.
Salado and Gila valleys, South Arizona, USA	B.C. 300 to A.D. 650	[HO, CR]			Earth and stone channels with plastered lining. Stepped profile.	Irrigations system built by the Hohokams. Re-used by the Mormons of Mesa City in late 19-th century.
Moche valley, Peru	A.D. 200-1500	[FA1, FA2]				Irrigation systems built by the Mochica civilisation, later extended by the Chimus and the Incas. Over 100 km of canals.
Cerro Orejas			0.012		Terrace irrigation system with free-falling nappes.	W = 0.2 to 0.3 m. h < 2 m.
Quishuarpata canal, Peru	AD 1000? to 1532	[FA1]	0.5		Canal floor made of irregularly shaped granite blocks with granite faced walls.	Artificial channel (6 km long) in the Hualancay river valley, near Cuzco. W ≤ 0.8 m.
Chute 1				23.3	Steep narrow steps. Drop structures near the end.	Pre-Inca design. 120 m long. h = 1 to 3 cm.
Chute 2A				32.6	Steep narrow steps.	Chutes 2A and 2B : 150 m long together. h = 1 to 3 cm.
Chute 2B				31	Steep narrow steps. Drop structures near the end.	h = 1 to 3 cm.
Cusichaca valley, Peru	A.D. 1200 to 1532	[FA1]			Terrace irrigation system with free-falling nappes and granite linings.	Inca irrigation system. h = 4 m.
Inca Irrigation systems, Peru Ollantaytambo	A.D. 1200 to 1532	[CO]		25 to 70 (a)	Terrace irrigation systems. Vertical channels and drop structures.	High altitude staircase farms (up to 3,400 m altitude). h = 2.4 to 4.3 m.
Vigevano, near La Sforzesca, Italy	A.D. 1500?	[RI]		26.6	Staircase waterfall (130 steps).	Described by LEONARDO DA VINCI. h = 0.152 m.

Notes : h : step height or drop height;

(a) : mountain slope

[CO] COOK (1916); [CR] CROUCH (1991); [FA1] FARRINGTON (1980); [FA2] FARRINGTON and PARK (1978); [HO] HODGE (1893); [LE] LeBARON BOWEN and ALBRIGHT (1958); [NE] NETZER (1983); [RI] RICHTER (1939).

2.4 Discussion

2.4.1 Dissemination of the spillway-design expertise

From the Antiquity up to the beginning of the 20-th century, the Romans, Moslems and Spanish contributed successively to the dissemination of the art of dam-building, including the expertise of stepped channel design. Dams and stepped overflow spillways were built very early in the Middle-East. Then the construction technique spread through the Mediterranean in Roman times. During their expansion period (7-th to 9-th centuries A.D.), the Moslems learned from the Sabaens, Nabataeans and Romans. The Muslim conquerors of the Hispanic peninsula brought their water traditions with them from the Eastern Mediterranean. Seven hundred years of Moorish control of Iberia left a strong hydraulic influence and a solid tradition of irrigation structures. Following the reconquest of Spain, the Catholic Spanish benefited from the Roman and Moslem precedents. Later, the Spanish conquered the New World and transferred deliberately their technology in turn.

Spain occupied a exceptional place in the development of large dams. Most European countries and European settlements in America benefited from their dam expertise, including in the field of stepped spillway design. Clear evidence of the Spanish influence were found certainly in France, Mexico and United States (table 2-3, fig. 2-2, 2-3 & 2-4).

It is worth noting the existence of some pre-European hydraulic expertise in Northern, Central and Southern America (e.g. SMITH 1971). Particularly, irrigation stepped channels were used in Arizona and Peru before the Spanish conquest (table 2-2).

In most early dams (table 2-1, fig. 2-2), the waters were discharged over the dam crests, and the stepped spillways were selected to contribute to the stability of the dam, for their simplicity of shape or for a combination of the two. Later, design engineers realised the advantages of stepped channels for energy dissipation purposes and to prevent scouring. By the fall of the 19-th century, overflow stepped spillways were selected frequently to contribute to the dam stability and to enhance energy dissipation (table 2-3, fig. 2-5). Most structures were masonry and concrete dams with a downstream stepped face reinforced by granite blocks. The spillway of the New Croton dam (1906) is probably the first stepped chute designed specifically to maximise energy dissipation (fig. 2-6).

In the first part of the 20-th century, new progress in the energy dissipation characteristics of hydraulic jumps (e.g. BAKHMETEFF and MATZKE 1936) favoured the design of stilling basins downstream of chutes and spillways. Stilling basins allowed larger energy dissipation and smaller structures. Altogether, the concept of downstream stilling basins contributed to cheaper hydraulic structure constructions.

Recently (in the 1970's), design engineers have regained interest for stepped spillways. The trend was initiated by the introduction of new construction materials : e.g. roller compacted concrete (RCC), strengthened-mesh gabions. Over the past decade, several dams have been built with overflow stepped spillway around the world (table 1-2).

Table 2-3 - Stepped spillways built between 1850 and 1950

Name	Year	Ref.	Dam height (m)	Slope α (deg.)	Construction	Comments
(1)	(2)	(3)	(4)	(5)	(6)	(7)
Nijar dam, Spain	1850	[SH, SM]	31		Curved gravity dam in masonry.	Overflow dam. Flat crest followed by 7 steep steps (h = 1.2 to 5.3 m).
Quinson dam (or Verdon dam), France	1870	[WE]	18	83 & 51	Gravity dam. Rubble with cut-stone facing.	Overflow dam (W = 40 m). Maximum discharge : 1,200 m³/s. Flat crest (4.6 m) followed by smooth slope (83 deg.) and 10 steps (51 deg., h = 0.75 m).
Hijar dam, Spain	1880	[SH]	43	53?	Curved masonry dam	Step height : h = 2 m.
Upper Barden reservoir, UK	1882	[BIA]	42		Earth embankment with masonry spillways at each end.	Maximum discharge : 43 m³/s. Flat steps at upstream end.
Arizona canal dam, USA	1887	[EN1]	10		Timber cribs filled with loose rock and gravel. Dam failure in 1905.	Maximum discharge : 33 m²/s. 3 horizontal steps.
Tytam dam, Hong Kong	1887	[SH]	29	65	Gravity dam (40% stone, 60% concrete).	Broad crest (6.4 m) followed by 9 steps (h = 3.05 m).
Castlewood dam, USA	1890	[SH]	21	45	Rockfill dam with masonry walls. Reinforced by an upstream earth embankment in 1900.	Maximum discharge : 115 m³/s. Broad crest (2.5 m) followed by 16 steps (h = 0.61 m). W = 30.5 m.
Goulburn weir, Australia (fig. 2-5)	1891	[EV]	15	~ 30	Concrete gravity weir with horizontal steps made of granite blocks.	Design discharge : 1,970 m³/s. Record discharge 1,982 m³/s on 7 June 1917. W = 141 m. 12 steps (h = 0.5 m).
Titicus dam, USA	1895	[WE]	41	~ 63	Earth dam with masonry overflow spillway (masonry of rubble with cut stone laid in regular courses).	W = 61 m. 13 steps (h = 0.61 to 3.7 m).
Canyon Ferry dam, USA	1898	[WE]	8.8		Timber cribs filled with stones. Reinforced by upstream earth embankment	W = 148 m. 3 steps (h = 2.3 to 3.05 m).
Yuba River debris barrier, USA	1904	[EN2]	6.8	15.4	Debris dam : hydraulic fill anchored by rockfill, piling and concrete facing.	Maximum discharge : 9.8 m²/s. W = 380 m. Smooth profiled steps with concrete facing. h = 1.8 to 2.6 m.
Pedlar river dam, USA	1905	[SH]	22.4	62	Concrete dam with spillway steps of granite blocks laid in cement mortar.	Maximum discharge : 7.7 m²/s. W = 45.7 m. 10 steps (h = 0.76 to 2.3 m). Ventilated steps.
Urft dam, Germany	1905	[KE, SH]	58		Curved gravity dam (slate masonry) with upstream earth embankment.	Maximum discharge : 200 m³/s. W = 100 m. Steps cut in natural rock covered by concrete (h = 1.52 m).
New Croton dam, USA (fig. 2-6)	1906	[W7, GO]	90.5	53	Masonry gravity dam. Spillway damage in 1955.	Maximum discharge : 1,550 m³/s. W = 305 m. h = 2.13 m.

Table 2-3 - Stepped spillways built between 1850 and 1950

Name	Year	Ref.	Dam height (m)	Slope α (deg.)	Construction	Comments
(1)	(2)	(3)	(4)	(5)	(6)	(7)
Wachusett dam, USA	1906	[WE]	62.5	18.4	Rubble masonry gravity dam. Separate overflow weir with granite facing.	Maximum discharge : 1,135 m^3/s. W = 137.2 m. h = 0.61 m.
Croton Falls dam, USA	1911	[WE]	52.7		Cyclopean masonry dam with concrete block facing. Horizontal steps made of granite.	Overflow weir : W = 213 m. 12 steps with rounded step edges : h = 0.61 m, l = 0.305 to 0.91 m.
Feather river diversion weir, USA	1912	[ET]	7.9	26.6	Crib weir with horizontal lumber steps.	Maximum discharge : 73 m^2/s. W = 85.3 m.
Lahontan dam, USA	1915	[RH]	49			Maximum discharge : 742 m^3/s.
Warren dam, Australia	1916	[JO]	17.4 (^{a})	35 (^{a})	Concrete gravity dam. Overtopped in 1917. Spillway re-designed in 1926.	Maximum spillway discharge : 100 m^3/s. W = 35 m. h = 0.37 m.
Hetch-Hetchy, USA	1923	[KE]	69.4 (^{a})	57 (^{a})	Gravity dam (granite masonry). Dam heightening in 1937.	Uncontrolled siphon-spillway followed by stepped chute (h = 2 m).
Gilboa dam, USA	1926	[GM]	49		Masonry dam	Maximum discharge : 3,130 m^3/s. Inclined downwards steps (δ = -5.7 deg.) : h = 6.1 m. W = 403.6 m.
St. Francis dam, USA	1926	[JA]	62.5		Concrete gravity dam. Dam failure in 1928.	Step height : h = 0.4 m. W = 67 m.
Eildon Weir, Australia	1927	[KN]	42.7 (^{a})	56 and 22 (^{a})	Rockfill dam with concrete spillway section. Spillway re-designed in 1936. New dam in 1955.	Maximum discharge : 566 m^3/s. Step height : h = 1.83 m. 14 steps (56 deg.) followed by a flat smooth slope and a 22-degree stepped channel. W = 208 m.
Paderno power plant, Austria	19??	[SC]	18	17.7	Pooled steps.	11 step cascade. Step height : h = 1.6 m. Pool height : d_t = 1 m.

Notes :

(^{a}) :   original dam

[BIA] BINNIE (1913); [EN1] Engineering News (1905a); [EN2] Engineering News (1905b); [ET] ETCHEVERRY (1916); [EV] EVANS (1984); [GM] GAUSMANN and MADDEN (1923); [GO] GOUBET (1992); [JA] JANSEN (1983); [JO] JOHNSON and TEMPLAR (1976); [KE] KELEN (1933); [KN] KNIGHT (1938); [RH] RHONE (1990); [SC] SCHOKLITSCH (1937); [SH] SCHUYLER (1909); [SM] SMITH (1971); [W7] WEGMANN (1907); [WE] WEGMANN (1911).

Fig. 2-5 - Goulburn weir (Australia, 1891) : view from the right bank (Courtesy of the Victoria Rural Water Corporation) - Photograph taken after the replacement of the gates in 1986

Fig. 2-6 - New Croton dam (USA, 1906) : view of the spillway from downstream, about 1910 (Courtesy of the American Society of Civil Engineers) - Note the masonry dam on the right, and the bridge connecting the dam crest to the right bank over the spillway weir

Fig. 2-7 - Staircase cascade by LEONARDO DA VINCI (after RICHTER 1939)

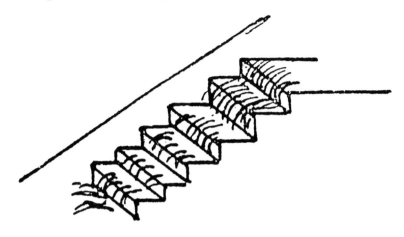

2.4.2 Design techniques

Since the Antiquity, the design of stepped spillways and channels was recognised to reduce flow velocities and to enhance energy dissipation. During the Renaissance period, Spanish and Italian hydraulicians (e.g. Juanelo TURRIANO, LEONARDO DA VINCI) applied these principles to staircase weirs. LEONARDO DA VINCI himself realised that the flow, "the more rapid it is, the more it wears away its channel"; if a waterfall "is made in deep and wide steps, after the manner of stairs, the waters (...) can no longer descend with a blow of too great a force" (fig. 2-7). He illustrated his conclusion with the staircase water falls at Vigevano, near La Sforzesca, Italy (130 steps, h = 0.15 m, h/l = 0.5) "down which the water falls so as not to wear away anything" (RICHTER 1939).

Some ancient engineers might have known the concepts of nappe and skimming flows. But there is evidence that, even at the beginning of the 20-th century, hydraulic engineers had no quantitative information on the main flow properties (e.g. flow resistance, head loss). It is only recently that new progress on the hydraulics of stepped channels has been achieved : e.g. ESSERY and HORNER (1978), SORENSEN (1985), RAJARATNAM (1990), PEYRAS et al. (1991).

The author notes with interest a continuity in the design of stepped spillways. The characteristics of ancient stepped spillways are similar and comparable to the present designs (fig. 2-8). Figure 2-8 presents the step height h, the maximum discharge capacity q_w and the ratio d_c/h for a wide range of dams as a function of year of completion. Since the Antiquity, the step heights increased slightly up to 1930 (fig. 2-8(A)). Recently, unlined rock spillways have been designed with larger step heights : i.e., 10-m high typically. But the most recent concrete and gabion structures (table 1-2) use smaller step heights, ranging from 0.2 to 1 m.

Surprisingly, most contemporary stepped spillways are designed for a maximum discharge capacity no greater than some ancient designs (fig. 2-8(B)). New stepped spillway designs might be cheaper to build but they do not provide necessarily larger discharge capacities !

In the past, most small weirs and drop structures were designed for a nappe flow regime. But figure 2-8(C) suggests interestingly that some ancient stepped spillways were designed for a skimming flow regime : i.e., $d_c/h > 0.4$ to 0.8 (see paragraph 4.2). On case, the Mestella weir (A.D. 960) was able to sustain very large skimming flows : i.e., q_w up to 85 m^2/s and d_c/h up to 26.

Stepped weirs and channels have been used fore more than 2,500 years. Despite recent advances in technology, the design characteristics (geometry, discharge) of stepped spillways show some continuity from the Antiquity up to now (fig. 2-8). Albeit the recent regain of interest for stepped spillways, the concept of "stepped chute" is not a new technique (nor a revolution) but barely an evolution of design !

Fig. 2-8 - Evolution of stepped spillway design over the ages

(A) Step height h

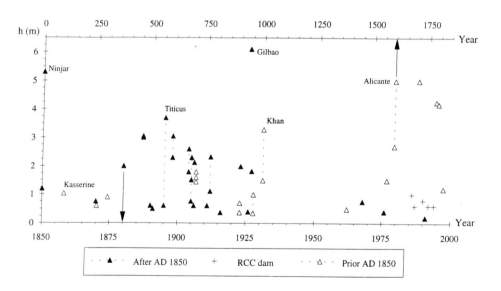

Fig. 2-8 - Evolution of stepped spillway design over the ages

(B) Maximum discharge capacity q_w

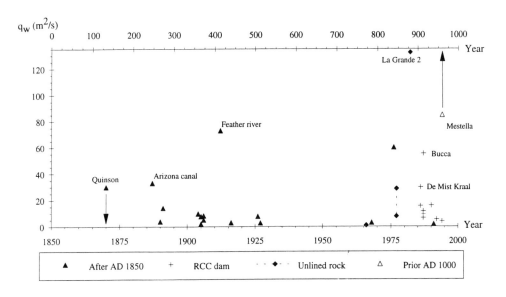

(C) Ratio d_c/h

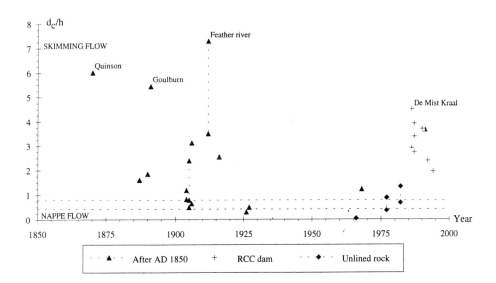

CHAPTER 3
NAPPE FLOW REGIME

3.1 Introduction

The nappe flow regime is defined as a succession of free-falling nappes. The flowing waters bounce from one step to the next one as a series of small free falls (fig. 3-1).

Three types of nappe flow can be distinguished : 1- nappe flow with fully-developed hydraulic jump for low flow rate and small flow depth (sub-regime NA1), 2- nappe flow with partially-developed hydraulic jump (sub-regime NA2), and 3- nappe flow without hydraulic jump (sub-regime NA3) (fig. 3-2). Nappe flows on horizontal steps are characterised typically by the presence of hydraulic jumps (sub-regimes NA1 and NA2). A nappe flow without hydraulic jump might occur for relatively large discharges, before the apparition of skimming flow. This sub-regime NA3 is observed more often on steep chutes and on channels with inclined-downwards steps (fig. 3-2(C)).

Along a chute with horizontal steps, a typical nappe flow situation consists of a series of free-fall jets impinging on the next step and followed by a hydraulic jump (fig. 3-2). The flow energy is dissipated by jet breakup in air, by jet impact and mixing, on the step and by the formation of a hydraulic jump on the step. Stepped channels with nappe flows can be analysed as a succession of drop structures.

3.2 Hydraulic characteristics of nappe flows

MOORE (1943) and RAND (1955) analysed a single-step drop structure (e.g. fig. 3-3). For a horizontal step, the flow conditions near the end of the step change from subcritical to critical at some section a short distance back from the edge. The flow depth at the brink of the step d_b is :

$$d_b = 0.715 * d_c \qquad (3-1)$$

where d_c is the critical flow depth (ROUSE 1936). Downstream of the brink, the nappe trajectory can be computed using potential flow calculations, complex numerical methods or approximate methods as that developed by MONTES (1992).

Application of the momentum equation to the base of the overfall leads to (WHITE 1943) :

$$\frac{d_1}{d_c} = \frac{2^{1/2}}{\frac{3}{2^{3/2}} + \sqrt{\frac{3}{2} + \frac{h}{d_c}}} \qquad (3-2a)$$

where d_1 is the flow depth at section 1 (fig. 3-4), and h is the step height. The total head H_1 at section 1 can be expressed non-dimensionally as :

$$\frac{H_1}{d_c} = \frac{d_1}{d_c} + \frac{1}{2} * \left(\frac{d_c}{d_1}\right)^2 \qquad (3-3)$$

Fig. 3-1 - Examples of nappe flow regime - Detail of the flow above stepped channels
(A) Peak tramway fountain, Hong Kong in 1993 - α = 20 deg., h = 0.15 m, horizontal steps
Flow from the right to the left (sub-regime NA1)

(B) Brushes Clough spillway, UK in 1994 (Courtesy of Mr GARDINER, NWW) - α = 18.4 deg., h
= 0.19 m, inclined steps (δ = -5.6 deg.) - Flow from the right to the left (sub-regime NA1)

The flow depth and total head at section 2 (fig. 3-4) are given by the classical hydraulic jump equations :

$$\frac{d_2}{d_1} = \frac{1}{2} * \left(\sqrt{1 + 8*Fr_1{}^2} - 1 \right)$$

(3-4)

$$\frac{H_1 - H_2}{d_c} = \frac{(d_2 - d_1)^3}{4 * d_1 * d_2 * d_c}$$

(3-5)

where Fr_1 is the Froude number defined at section 1 : $Fr_1 = q_w / \sqrt{g * d_1{}^3}$.

RAND (1955) re-analysed several experiments and developed the following correlations :

Fig. 3-2 - Nappe flow sub-regimes

(A) Nappe flow with fully-developed hydraulic jump (sub-regime NA1)

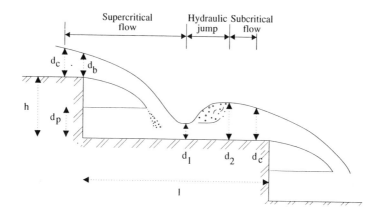

(B) Nappe flow with partially-developed hydraulic jump (sub-regime NA2)

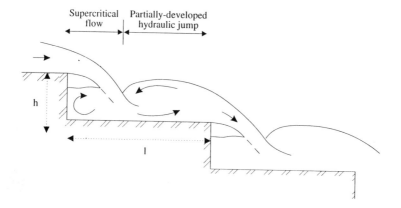

Fig. 3-2 - Nappe flow sub-regimes

(C) Nappe flow without hydraulic jump (sub-regime NA3)

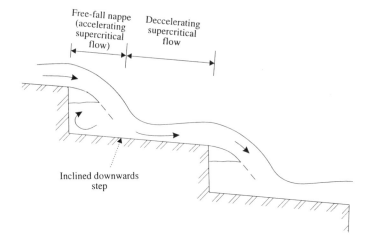

Fig. 3-3 - Gabion drop structure near Pisino, Croatia (former Yugoslavia) (1974) (Courtesy of Officine Maccaferri) - Note the downstream counterweir and the concrete linings protecting both the drop crest and counterweir

Fig. 3-4 - Flow at a drop structure

(A) Flow at a drop structure

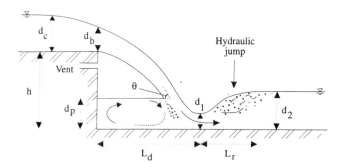

(B) Detail of the nappe geometry

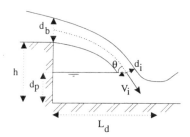

$$\frac{d_1}{h} = 0.54 * \left(\frac{d_c}{h}\right)^{1.275} \tag{3-6}$$

$$\frac{d_2}{h} = 1.66 * \left(\frac{d_c}{h}\right)^{0.81} \tag{3-7}$$

$$\frac{d_p}{h} = \left(\frac{d_c}{h}\right)^{0.66} \tag{3-8}$$

$$\frac{L_d}{h} = 4.30 * \left(\frac{d_c}{h}\right)^{0.81} \tag{3-9}$$

where d_p is the height of water in the pool behind the overfalling jet, L_d is the distance from the drop wall to the position of the depth d_1 (fig. 3-4). Equation (3-6) is an empirical correlation that fits well equation (3-2) (appendix A). Note also that equations (3-6) to (3-9) were obtained for an aerated nappe. Flow properties for un-aerated nappes are summarised in appendix A.

The flow conditions at the impact of the nappe with the receiving pool can be deduced from the equation of motion. Using equations (3-1) and (3-8), the nappe thickness d_i, the nappe velocity V_i and the angle θ of the nappe with the horizontal, at the impact, can be correlated by :

$$\frac{d_i}{h} = 0.687 * \left(\frac{d_c}{h}\right)^{1.483} \tag{3-10}$$

$$\frac{V_i}{V_c} = 1.455 * \left(\frac{d_c}{h}\right)^{-0.483} \tag{3-11}$$

$$\tan\theta = 0.838 * \left(\frac{d_c}{h}\right)^{-0.586} \tag{3-12}$$

Downstream of the impact of the nappe, the roller length of a fully-developed hydraulic jump is estimated as (HAGER et al. 1990) :

$$\frac{L_r}{d_1} = 8 * \left(\left(\frac{d_c}{d_1}\right)^{3/2} - 1.5\right) \tag{3-13}$$

where L_r is the roller length and d_1 is the flow depth at the jump toe : i.e., at section 1 (fig. 3-4). If the length of the drop L_d plus the length of the roller L_r are smaller than the length of a step l, a fully developed hydraulic jump can take place (fig. 3-2 and 3-4). Combining equations (3-9) and (3-13), a condition for nappe flow regime with fully developed hydraulic jump (sub-regime NA1) is deduced. A nappe flow regime with fully-developed hydraulic jump occurs for discharges smaller than a critical value defined as :

$$\left(\frac{d_c}{h}\right)_{char} = 0.0916 * \left(\frac{h}{l}\right)^{-1.276} \tag{3-14}$$

where l is the step length. Nappe flow situations with fully-developed hydraulic jump occur for $d_c/h < (d_c/h)_{char}$. Equation (3-14) was obtained for : $0.2 \leq h/l \leq 6$. For steep slopes (i.e. $h/l > 0.5$), the sub-regime NA1 can occur only for very low flow rates (fig. 4-2).

Along a stepped chute, critical flow conditions take place next to the end of each step, and equations (3-2) to (3-13) provide the main flow parameters for a nappe flow regime with fully developed hydraulic jump (sub-regime NA1). PEYRAS et al. (1991, 1992) indicated that these equations can be applied also with reasonable accuracy to nappe flows with partially developed jumps (sub-regime NA2).

Discussion

The above calculations (eq. (3-1), (3-2), (3-6) to (3-12)) have been developed assuming that the flow upstream of the brink of the overfall is *subcritical*, and hence critical immediately upstream of the brink. This assumption is valid for an un-gated channel with horizontal steps and for nappe flow situations with hydraulic jump (sub-regimes NA1 and NA2). For a nappe flow without hydraulic jump, the waters flow as a supercritical stream and critical flow conditions are not observed at the step brink. The flow consists of a succession of free-falling nappes, jet impacts and decelerating supercritical flows (fig. 3-2(C)). Also, if the upstream end of the channel is controlled by an underflow gate (e.g. sluice or tainter gate), the flow at the brink of the first step might be supercritical.

In absence of critical flow conditions next to the step brink, the hydraulic characteristics of supercritical nappe flows are determined by the nappe trajectory, the jet impact on the step and the flow resistance on the step downstream of the nappe impact. Several researchers (ROUSE 1943, RAJARATNAM and MURALIDHAR 1968, HAGER 1983, MARCHI 1993) gave details of the brink flow characteristics and of jet shape for supercritical flows. For supercritical overfalls, the application of the momentum equation at the base of the overfall, using the same method as WHITE (1943), leads to the result :

$$\frac{d_1}{d_c} = \frac{2 * Fr^{-2/3}}{1 + \frac{2}{Fr^2} + \sqrt{1 + \frac{2}{Fr^2} * \left(1 + \frac{h}{d_c} * Fr^{2/3}\right)}} \qquad (3\text{-}2b)$$

where Fr is the Froude number of the supercritical flow upstream of the overfall brink.

3.3 Energy dissipation

In a nappe flow situation with a fully developed hydraulic jump, the head loss at any intermediary step equals the step height. The energy dissipation occurs by jet breakup and jet mixing, and with the formation of a hydraulic jump on the step. The total head loss along the chute ΔH equals the difference between the maximum head available H_{max} and the residual head at the downstream end of the channel H_1 (eq. (3-3)). In dimensionless form, it yields :

$$\frac{\Delta H}{H_{max}} = 1 - \left(\frac{\frac{d_1}{d_c} + \frac{1}{2} * \left(\frac{d_c}{d_1}\right)^2}{\frac{3}{2} + \frac{H_{dam}}{d_c}}\right) \qquad \text{Un-gated chute} \quad (3\text{-}15a)$$

$$\frac{\Delta H}{H_{max}} = 1 - \left(\frac{\frac{d_1}{d_c} + \frac{1}{2} * \left(\frac{d_c}{d_1}\right)^2}{\frac{H_{dam} + H_o}{d_c}}\right) \qquad \text{Gated chute} \quad (3\text{-}15b)$$

where H_{dam} is the dam crest head above downstream toe and H_o is the reservoir free-surface elevation above the chute crest. For an un-gated channel, the maximum head available and the dam height are related by : $H_{max} = H_{dam} + 1.5*d_c$. For a gated waterway : $H_{max} = H_{dam} + H_o$.

The residual energy is dissipated at the toe of the chute by a hydraulic jump in the dissipation basin. Combining equations (3-6) and (3-15), the total energy loss becomes :

$$\frac{\Delta H}{H_{max}} = 1 - \left(\frac{0.54 * \left(\frac{d_c}{h}\right)^{0.275} + \frac{3.43}{2} * \left(\frac{d_c}{h}\right)^{-0.55}}{\frac{3}{2} + \frac{H_{dam}}{d_c}}\right) \qquad \text{Un-gated chute} \quad (3\text{-}16a)$$

$$\frac{\Delta H}{H_{max}} = 1 - \left(\frac{0.54 * \left(\frac{d_c}{h}\right)^{0.275} + \frac{3.43}{2} * \left(\frac{d_c}{h}\right)^{-0.55}}{\frac{H_{dam} + H_0}{d_c}} \right) \qquad \text{Gated chute} \quad (3\text{-}16b)$$

On figure 3-5, the head loss for an un-gated structure (eq. 3-16a) is plotted as a function of the critical flow depth and the number of steps, and compared with experimental data (MOORE 1943, RAND 1955, HORNER 1969, STEPHENSON 1979a). Figure 3-5 indicates that most of the flow energy is dissipated on the stepped channel for large dams (i.e. large number of steps). Further, for a given dam height, the rate of energy dissipation decreases when the discharge increases. Note the good agreement between equation (3-16a) and model data obtained on drop structures (1 step) and multiple-step channels (fig. 3-5).

Equations (3-15) and (3-16) were obtained for nappe flows with fully developed hydraulic jumps (sub-regime NA1). PEYRAS et al. (1991) performed experiments for nappe flows with fully and partially developed hydraulic jumps. The rate of energy dissipation of nappe flows with partially developed hydraulic jumps (sub-regime NA2) was within 10% of the values obtained for nappe flows with fully developed hydraulic jump for similar flow conditions. Therefore, it is believed that equation (3-16) may be applied with a reasonable accuracy to most nappe flow situations on horizontal-step chutes.

Fig. 3-5 - Energy dissipation in nappe flow regime - Comparison between equation (3-16) and experimental data (MOORE 1943, RAND 1955, HORNER 1969, STEPHENSON 1979a)

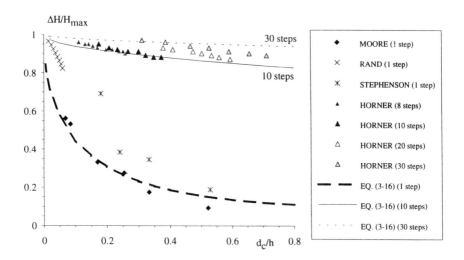

Fig. 3-6 - Nappe flow with pooled steps

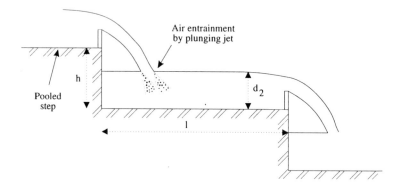

3.4 Air entrainment in nappe flow regime

3.4.1 Introduction

In a nappe flow regime, air is entrained at the intersection of the falling nappe with the receiving pool and in the hydraulic jump downstream of the impact of the nappe (fig. 3-2). The first process is called a plunging jet entrainment. For deep pooled steps (fig. 3-6), air is entrained only at the intersection of the nappe with the receiving pool of water.

In the first part of this section, the nappe ventilation of a free-falling nappe is discussed. Then the plunging entrainment process is reviewed. Later the mechanisms of air entrainment by hydraulic jump are summarised. The results are then applied to nappe flow situations.

It must be noted that, for large step heights (i.e. h > 5 m), air entrainment may occur along the upper and lower air-water interfaces of the falling nappe. Experiments were performed by ERVINE and FALVEY (1987) on circular jets and by CHANSON (1993b) with two-dimensional jets. These studies may provide useful information on the amount of air entrained along the interface. In most practical situations however, the effects of air entrainment along the nappe free-surfaces are small and can be neglected.

3.4.2 Nappe ventilation and oscillations

Nappe ventilation

When the falling nappe intersects the receiving pool of water, air is entrained within the flow. The air entrained at the intersection between the pool and the underside of the jet is drawn from the air cavity beneath the nappe. If the cavity between the nappe and the vertical step is not ventilated, the pressure in the cavity falls below atmospheric, with subpressure and nappe oscillations occurring. LEVIN (1968, pp. 28-37) gave indications of appropriate levels of nappe ventilation. The re-analysis of his data indicates that the nappe ventilation is estimated as :

$$\frac{Q_{air}^{nappe}}{Q_w} = 0.19 * \left(\frac{h - d_p}{d_b}\right)^{0.95} \qquad \text{for } 3 < Fr < 10 \quad (3\text{-}17a)$$

$$\frac{Q_{air}^{nappe}}{Q_w} = 0.21 * \left(\frac{h - d_p}{d_b}\right)^{1.03} \qquad \text{for } 13 < Fr < 15 \quad (3\text{-}17b)$$

where Q_{air}^{nappe} is the nappe aeration and Fr is a Froude number defined in term of the nappe thickness. Note that equations (3-17a) and (3-17b) apply to both bottom outlet aeration and nappe ventilation. For the ventilation of free-falling nappes, LEVIN (1968) recommended the use of equation (3-17a). Figure 3-7 shows a comparison between equations (3-17) and LEVIN's (1968) data.

For a wide stepped spillway (Pedlar river dam, fig. 2-2(G), table 2-3), SCHUYLER (1909) described a ancient ventilation system : each step was ventilated by a 0.15-m diameter pipe open at each sidewall end and three open ends extending to the vertical face of the step. Note that, at the Pedlar river spillway, the maximum discharge implied a maximum critical depth over step height d_c/h of about 0.8 to 2.4 (i.e skimming flow regime, fig. 4-3). The ventilation system was designed probably for flow rates smaller than the maximum discharge.

Fig. 3-7 - Nappe ventilation

Comparison between equation (3-17) and experimental data (LEVIN 1968)

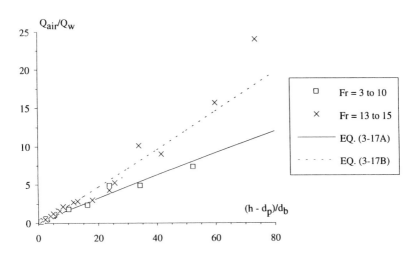

Fig. 3-8 - Free-falling nappe oscillations

Comparison between equation (3-18) and experimental data (table 3-1)

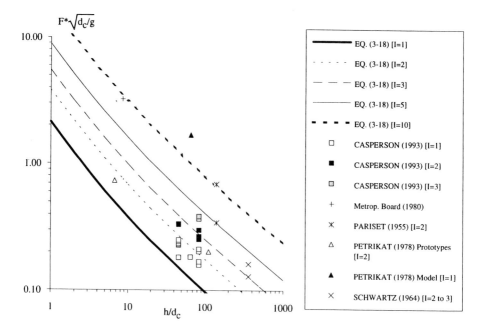

Nappe oscillations

Thin nappes of liquid discharging over long free-flow spillways can be subject to nappe oscillations. These fluttering instabilities, also called Kelvin-Helmholtz instabilities, induce oscillations of the water sheet and air movement below the nappe strong enough to be heard miles away (PARISET 1955, PETRIKAT 1958, THOMAS 1976, CASPERSON 1993). The danger to the structure is relatively small but the associated noise is a nuisance.

Nappe oscillations are controlled by the motion of the air trapped in the space beneath the nappe. The natural frequency of the air-water system depends on the mass of water and on the volume of trapped air. To a first approximation, the possible frequencies of oscillation F of a nappe leaving an horizontal step can be estimated as (appendix A) :

$$\frac{F}{\sqrt{\frac{g}{d_c}}} = 0.715 * \frac{I + \frac{1}{4}}{\left(-1 + \sqrt{1 + 1.022 * \frac{h}{d_c}}\right)} \tag{3-18}$$

where h is the step height, and I is an integer : I = 1, 2, 3, 4, 5 or more. For an oscillating nappe, the number of wavelengths in the nappe approximates the integer I. On figure 3-8, equation (3-

18) is compared with model and prototype observations (table 3-1).

The occurrence of nappe oscillations is a function of the mass of the free-falling nappe, the length of the nappe and the volume of air trapped beneath the nappe (e.g. PETRIKAT 1978), CASPERSON 1993). Based on model studies, SCHWARTZ (1964a) indicated that the fluttering instabilities can occur for discharges smaller than 2.1 m^2/s.

PETRIKAT (1958,1978) and SCHWARTZ (1964a) suggested the use of nappe splitters to control the oscillation of the nappe. Splitters are placed on the crest at regular intervals. The addition of splitters creates numerous gaps in the nappe, allowing the passage for air. Since only thin nappes oscillate, the splitters need to be between 0.2 to 0.75 m high and the maximum spacing between splitters must be no greater than two-thirds of the fall height (2/3*h). Splitters were successfully installed at several overfall dams in Australia and New Zealand (e.g. Repulse, Scrivener, Devils Gate, Waitaki dams) as illustrated by THOMAS (1976).

Note that splitters are used also to enhance energy dissipation of large free-falling nappes. In South Africa, they are in use since 1938 : e.g. Loskop dam, P.K. Le Roux dam (BACK et al. 1973, ROBERTS 1977).

Other methods to prevent nappe instabilities include modifications of the crest profile or a roughening of the crest (e.g. addition of stones to the crest at Avon dam, Metropolitan Water 1980).

In most nappe flow situations, the falling nappe is short, and, if the ventilation is adequate, the risk of nappe oscillation and associated noise is small.

Table 3-1 - Observations of free-falling nappe oscillations

Reference	q_w	h	F	I	Nb of Exp.	Remarks
	m^2/s	m	Hz			
(1)	(2)	(3)	(4)	(5)	(6)	(7)
PARISET (1955)	0.0066	1.65	20	2		Model data. W = 2.75 m. 0.03 < d_t < 0.11 m.
			10			0.22 < d_t < 0.5 m.
SCHWARTZ (1964a)	0.0023	0.9	10	3	1	Model data. Fig. 8.
			8	4	1	Fig. 9.
PETRIKAT (1978)	0.0041	0.797	48	1		Model data. W = 0.6 m. 0 < d_t < 0.41 m.
	0.0205	2.8	4	2	1	Prototype data. W = 30 m.
	0.6028	2.25	3 to 5	2		Upper weir at Hameln (Ger).
Metropolitan Board (1980)	0.025 to 0.034	0.38	48			Model data. W = 0.61 m.
CASPERSON (1993)	0.0075	1.175	4.3	1	1	Dunedin fountain No. 1 (NZ). W = 6 m.
	0.003 to 0.0075	0.825	4.2 to 6.7	1	12	Dunedin fountain No. 2 (NZ). W = 8 m.
			7.9 to 9.5	2	5	
			11.7 to 12.3	3	2	

Notes :

d_t	tailwater depth
I :	number of wavelengths in the nappe
Nb of Exp. :	number of experimental data
W :	channel (or crest) width

3.4.3 Air entrainment by plunging nappe

When the falling nappe impinges the receiving pool of water, air bubbles are entrained at the intersection of the jet with the receiving pool (fig. 3-4 and 3-6). This process is called plunging jet entrainment. Typical examples are shown on figure 3-9. Recent reviews on plunging jet entrainment include WOOD (1991), CHANSON and CUMMINGS (1992) and BIN (1993).

Air entrainment by two-dimensional jets (e.g. free-falling nappe) is caused by vortices with axes perpendicular to the flow direction. Several studies (tables 3-1 and 3-2) showed that the quantity of air entrained can be estimated as :

$$Q_{air}^{jet} = K' * (V - V_e)^n \qquad (3-19)$$

Fig. 3-9 - Examples of plunging jet flow situations

Vertical plunging jet

ERVINE and ELSAWY (1975)
ERVINE et al. (1980)
ERVINE and AHMED (1982)
KOGA (1982)
SENE (1984-88)

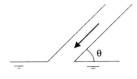

Inclined plunging jet

VAN DE SANDE and SMITH (1972-73-76)
ERVINE and ELSAWY (1975)
KOGA (1982)
SENE (1984-88)
DETSCH and SHARMA (1990)
KUSABIRAKI et al. (1992)

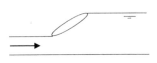

Hydraulic jump

KALINSKE and ROBERTSON (1943)
RAJARATNAM (1962)
WISNER (1965)
CASTELEYN et al. (1977)
(hydraulic jump in siphon)

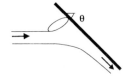

Impingement of a free jet
on an inclined wall

RENNER (1973-75)

Fig. 3-10 - Inception velocity for plunging jet V_e as a function of the jet turbulent intensity Tu for vertical circular jets (McKEOGH 1978, ERVINE et al. 1980) and vertical two-dimensional jets (CUMMINGS 1994)

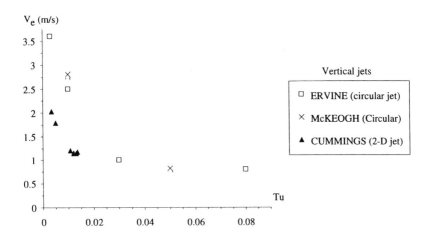

where K' is a constant, V is the jet velocity and V_e is the velocity at which air entrainment commences. Dimensional analysis indicates that the onset velocity V_e for plunging jet entrainment is given by :

$$V_e = F_1(\rho_w ; \mu_w ; \sigma ; g ; u_t ; L_t ; \theta) \tag{3-20}$$

where u_t is a measure of the turbulent velocity fluctuation, L_t is a measure of the turbulent length scale and θ is the angle of the jet with the liquid free-surface (fig. 3-4). In dimensionless terms, equation (3-20) becomes :

$$\frac{V_e}{\left(\dfrac{\mu_w^3 * g}{\sigma^2 * \rho_w}\right)} = F_2\left(Z ; \frac{u_t}{\left(\dfrac{\mu_w^3 * g}{\sigma^2 * \rho_w}\right)} ; \frac{L_t}{\left(\dfrac{\sigma^2}{\mu_w^2 * g}\right)} ; \theta\right) \tag{3-21}$$

where $\{\sigma^2/(\mu_w^2 * g)\}$ is a length scale based on the fluid properties and the gravity acceleration, $\{\mu_w^3 * g/(\sigma^2 * \rho_w)\}$ is a velocity scale based on the fluid parameters, and Z is the Morton number also called the liquid parameter and defined as :

$$Z = \frac{g * \mu_w^4}{\rho_w * \sigma^3}$$

Most experimental data show that the onset velocity V_e does not depend on the diameter or the thickness of the jet. Further, experimental results obtained by ERVINE et al. (1980) on vertical circular jets suggest that the inception velocity V_e is almost constant for large turbulent intensities (i.e. Tu > 3%), with typical values ranging from 0.8 to 1 m/s (table 3-2, fig. 3-10). But recent experiments (CUMMINGS 1994) suggest that V_e might be smaller for two-dimensional vertical jets (fig. 3-10).

Table 3-2 - Measurements of inception velocity for plunging jets

Reference	Geometry	Tu (%)	V_e (m/s)	Comments
(1)	(2)	(3)	(4)	(5)
LIN and DONNELLY (1966)	Vertical circular jet		$\dfrac{1.4886 * \sigma^{0.794}}{\rho_w^{0.206} * \mu_w^{0.587} * d^{0.206}}$	$4 < d < 8.1$ mm. Liquids : water, oil, glycerol.
			$\dfrac{2.60}{d^{0.206}}$	Air and water at 20 Celsius.
ERVINE & ELSAWY (1975)	Two-dimensional vertical jet		1.1	
McKEOGH (1978)	Circular vertical jet	1 5	2.8 0.8	$2.75 < d < 14.5$ mm.
LARA (1979)	Circular vertical jet		0.7 to 4	V_e varies with the jet length.
ERVINE et al. (1980)	Circular vertical jet	0.3 1 3 8	3.6 2.5 1.0 0.8	
ERVINE & AHMED (1982)	Two-dimensional vertical dropshaft		0.8	
KOGA (1982)	Circular jet		$2.58*\theta - 0.30$	$d = 4.6$ mm $\pi/7.2 < \theta < \pi/2.8$
	Rectangular jet		$1.73*\theta - 0.73$	$d = 1.2$ mm $\pi/3 < \theta < \pi/2$
SENE (1984)	Two-dimensional nozzle jet		1	
	Two-dimensional supported jet		1	Rough turbulent flow.
			2	Potential jet flow.
DETSCH and STONE (1992)	Circular jet		$4.25\text{E-}5 * \dfrac{\rho_w*\mu_w}{\sigma} * \exp(4.383*\theta)$	Liquid : water, saltwater, ethanol, ethylene glycol solutions. $d = 1.5$ mm. $\pi/12 < \theta < \pi/3$
			$5.79\text{E-}4 * \exp(4.383*\theta)$	Air and water at 20 Celsius. $\pi/12 < \theta < \pi/3$
CUMMINGS (1994)	Two-dimensional vertical supported jet	0.34 0.5 1.08 1.22 1.32	2.03 1.79 1.20 1.15 1.14	$d = 5$ to 15 mm.
WOOD (1991)	Circular vertical jet		$\dfrac{V_e^3}{g*\dfrac{\mu_w}{\rho_w}} = 0.5$ to $1\ 10^5$	Dimensional analysis.

Note :　θ : jet angle in radians

Table 3-3 - Quantity of air entrained by plunging jets

Reference	Geometry	Q_{air}^{jet}/Q_w	Comments
(1)	(2)	(3)	(4)
LIN and DONNELLY (1966)	Vertical circular jet	$K' * (V - V_e)$	$1 < V < 2.1$ m/s. $d = 5$ mm.
VAN DE SANDE and SMITH (1972)	Circular jet	$q_{air}^{jet} = K' * (\sin\theta)^{-1.2}$	
VAN DE SANDE and SMITH (1973)	Circular jet	$q_{air}^{jet} = K' * (\sin\theta)^{-1.4}$	$30 < \theta < 75$ degrees
		$q_{air}^{jet} = K' * V^{2.6}$	$V < 5$ m/s $\theta = 30$ degrees
		$q_{air}^{jet} = K' * V^{0.53}$	$5 < V < 10$ m/s $\theta = 30$ degrees
		$q_{air}^{jet} = K' * V^{1.8}$	$10 < V < 25$ m/s $\theta = 30$ degrees
ERVINE & ELSAWY (1975)	Two-dimensional jet	$K' * \left(1 - \dfrac{V_e}{V}\right)$	$1.5 < V < 9$ m/s
RENNER (1975)	Rectangular jet impinging horizontally	$K'(\theta) * Fr^{2.0}$	$2 < Fr < 9$ $30 < \theta < 90$ degrees
	onto a solid wall	$K'(\theta) * Fr^{0.77}$	$9 < Fr$ $30 < \theta < 90$ degrees
VAN DE SANDE and SMITH (1976)	Low velocity circular jet	$Q_{air}^{jet} = K' * \dfrac{d^{3/2} * V^{9/4}}{(\sin\theta)^{9/8}}$	$2 < V < 5$ m/s $2.8 < d < 10$ mm $20 < \theta < 60$ degrees Shorts jets.
McKEOGH (1978)	Vertical circular jet	$q_{air}^{jet} = 0.0018*d*(V-V_e)^{1.5}$	$V < 9$ m/s $2.75 < d < 14.5$ mm
YAGASAKI and KUZUOKA (1979)	Vertical circular jet	$q_{air}^{jet} = K' * V^{3.2}$	$V < 4$ to 7 m/s $2.8 < d < 7.3$ mm
		$q_{air}^{jet} = K' * V^{1.4}$	$7 < V < 12.5$ m/s
ERVINE & AHMED (1982)	Two-dimensional vertical dropshaft	$q_{air}^{jet} = 0.00045*(V - V_e)^3$	$\theta = 90$ degrees $3 < V < 6$ m/s
SENE (1988)	Supported two-dimensional jet	$0.0055*Fr^2$	Smooth and rough turbulent jets : $3 < Fr < 5.5$.
		$0.0022*Fr^2$	Rough turbulent jets : $5.5 < Fr < 8$.
SENE (1988)	Low velocity jet	$K' * Fr^2$	Dimensional analysis
	High velocity jet	$q_{air}^{jet} = K' * V^{3/2}$	Dimensional analysis
WOOD (1991)	High velocity jet	$K'(Re,Tu) * Fr^2$	Dimensional analysis

Note : V_e critical velocity (table 3-2)

Experimental data obtained with inclined jets (KOGA 1982, DETSCH and STONE 1992) indicate that the critical velocity V_e decreases when the angle of the jet decreases (table 3-2). It must be emphasised that these results were obtained only with circular jets of small dimensions.

Fig. 3-11 - Mechanisms of air entrainment by plunging jets

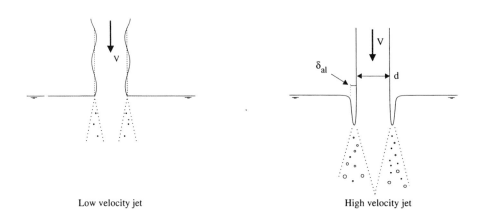

Low velocity jet   High velocity jet

At low jet velocities, air bubble entrainment is caused by the plunge pool water being unable to follow the undulations of the jet surface and small air pockets are formed. Air enters the flow following the passage of these disturbances through the interface between the jet and the receiving flow (fig. 3-11). The entrainment process is intermittent and pulsating. At high jet velocities (i.e. V > 8 to 12 m/s), experiments on circular plunging jets (VAN DE SANDE and SMITH 1973) indicate a qualitative change in the air entrainment process. An air layer, set into motion by shear forces at the surface of the jet, enters the flow at the impact point. This air layer becomes established and is continuous along the interface between the jet and the pool water (fig. 3-11). At low and high velocities, the air bubble transport is attributed to an entrapment phenomenon by large scale transient eddies which carry away air bubbles (GOLDRING et al. 1980, THOMAS et al. 1983).

A re-analysis of previous works (table 3-3) shows that the dimensionless quantity of air entrained can be estimated as :

$$\frac{Q_{air}^{jet}}{Q_w} = k_1 * Fr^2 \qquad\qquad\qquad\qquad \text{for V < 5 m/s (3-22a)}$$

$$\frac{Q_{air}^{jet}}{Q_w} = k_2 * \frac{1}{\sqrt{Fr}} \qquad\qquad\qquad \text{for 5 < V < 10 m/s (3-22b)}$$

$$\frac{Q_{air}^{jet}}{Q_w} = k_3 * Fr \qquad\qquad\qquad\qquad \text{for V > 10 m/s (3-22c)}$$

where Fr is the jet Froude number defined as $Fr = (V-V_e)/\sqrt{g*d}$ and d is the jet thickness (for a plane jet) or the jet diameter (for a circular jet).

Equation (3-22) is in agreement with experimental observations indicating different jet behaviours between low velocity jets (eq. (3-22a)) and high velocity jets (eq. (3-22c)). In the transition region between laminar and turbulent regimes, DETSCH and SHARMA (1990) suggested the occurrence of an intermittent vortex mechanism.

For low velocity jets, the effects of the jet angle were investigated by VAN DE SANDE and SMITH (1976) and KUSABIRAKI et al. (1990) : the quantity of air entrained can be expressed as :

$$\frac{Q_{air}^{jet}}{Q_w} = k_4 * \frac{Fr^2}{(\sin\theta)^{1.2}} \tag{3-23}$$

For a stepped chute with nappe flow regime, the flow conditions at the impact of the nappe are deduced from equations (3-10) to (3-12). Equations (3-22) or (3-23) can provide a first estimate of the quantity of air entrained. In equation (3-23), the results of SENE (1988) can provide a first estimate of the constant of proportionality : that is, $k_4 = 0.0055$ for $V_i/\sqrt{g^*d_i} < 5.5$.

NAKASONE (1987) observed the aeration at the base of overfalls. For deep receiving pools, he correlated the horizontal length of the bubbly region as :

$$\frac{L_a}{d_c} = 31.43 * h^{0.134} \tag{3-24}$$

where L_a is the bubble zone length measured from the vertical face of the step and h is in metres. Note that equation (3-24) was obtained for step heights in the range 0.5 to 2.2 m and critical depths ranging from 0.024 to 0.383 m.

It must be emphasised that equations (3-22), (3-23) and (3-24) were obtained for deep receiving pools with no or slow flow motion in the receiving pool. Such conditions do not apply to most stepped channel flow conditions. For small discharges, the tailwater depth is shallow and several researchers indicated that the rate of aeration decreases with a decreasing pool depth (ERVINE and ELSAWY 1975, NAKASONE 1987). SENE (1984) suggested that the entrainment of bubbles tends to zero when the depth of flow becomes shallower than the characteristic length of the eddies entrapping the bubbles.

3.4.4 Air entrainment by hydraulic jump

In most nappe flow situations, a hydraulic jump takes place immediately downstream of the impact of the falling nappe (fig. 3-2) and additional air bubbles are entrained at the toe of the jump.

The hydraulic jump is a limiting case of a plunging jet situation. Large quantities of air are entrained at the jump toe into the free shear layer which is characterised by intensive turbulence production, and large vortices with horizontal axes perpendicular to the flow. Dimensional analysis suggests that the quantity of air entrained is expressed as (WOOD 1991) :

$$\frac{Q_{air}^{HJ}}{Q_w} = K_1 * Fr^2 \tag{3-25}$$

where Q_{air}^{HJ} is the quantity of air entrained, Q_w is the water discharge, K_1 is a constant, Fr is the Froude number defined as $Fr = (V_1-V_e)/\sqrt{g^*d_1}$ and V_e is the onset velocity of air entrainment for hydraulic jump (table 3-4). A summary of experimental data is presented in table 3-5.

In practice, the correlations of RAJARATNAM (1967) and WISNER (1965) are commonly used :

$$\frac{Q_{air}^{HJ}}{Q_w} = 0.018 * (Fr_1 - 1)^{1.245} \qquad \text{(RAJARATNAM 1967) (3-26)}$$

$$\frac{Q_{air}^{HJ}}{Q_w} = 0.014 * (Fr_1 - 1)^{1.4} \qquad \text{(WISNER 1965) (3-27)}$$

where Fr_1 is the upstream Froude number of the jump (i.e. $Fr_1 = q_w/\sqrt{g*d_1^3}$) and d_1 is the upstream flow depth. Both equations are compared with experimental data on figure 3-12.

HAGER (1992) re-analysed the original data of RAJARATNAM (1967). He showed that the aeration length of the jump can be estimated as :

$$\frac{L_a}{d_2} = 3.5 * \sqrt{Fr_1 - 1.5} \qquad (3-28)$$

where L_a is the aeration length and d_2 is the downstream flow depth. In general, the aeration length L_a is larger than the roller length L_r (eq. (3-13)).

Fig. 3-12 - Air entrainment in hydraulic jump

Comparison between equations (3-26) and (3-27), and experimental data (RAJARATNAM 1962, RESCH and LEUTHESSER 1972, SCHRÖDER 1963)

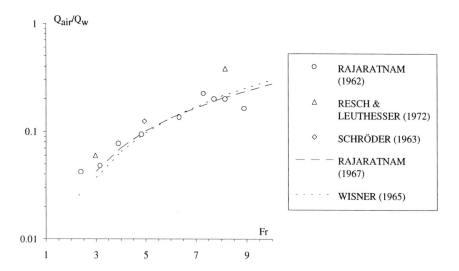

Table 3-4 - Inception velocity of air entrainment by hydraulic jumps

Reference (1)	Geometry (2)	V_e (m/s) (3)	Comments (4)
KALINSKE & ROBERTSON (1943)	Hydraulic jump in horizontal circular pipe	1.0	Model data.
WISNER (1965)	Hydraulic jump in rectangular conduit	$Fr_e = 1$	Prototype data.
CASTELEYN et al. (1977)	Siphon of square cross-section	0.8	Model data.
ERVINE and AHMED (1982)	Two-dimensional vertical dropshaft	0.8	Model data. Vertical jet.
RABBEN et al. (1983)	Hydraulic jump in horizontal pipes.	$Fr_e = 1$	Model data. Gate opening : 1/4.
CHANSON (1993e)	Hydraulic jump in rectangular channel	0.66 to 1.41	W = 0.3 m. Fully developed U/S shear flow.

Note : Fr_e :   Froude number defined in term of the inception velocity : $Fr_e = V_e/\sqrt{g^*d_1}$

Table 3-5 - Quantity of air entrained by hydraulic jumps

Reference (1)	Geometry (2)	Q_{air}^{HJ}/Q_w (3)	Comments (4)
KALINSKE & ROBERTSON (1943)	Hydraulic jump in horizontal circular pipe	$0.0066*(Fr_1 - 1)^{1.4}$	Model data. $2 < Fr_1 < 25$
WISNER (1965)	Hydraulic jump in rectangular conduit	$0.014*(Fr_1 - 1)^{1.4}$	Prototype data. $5 < Fr_1 < 25$
RAJARATNAM (1967)	Hydraulic jump in rectangular channel	$0.018*(Fr_1 - 1)^{1.245}$	Model data. $2.4 < Fr_1 < 8.7$
CASTELEYN et al. (1977)	Siphon of square cross-section	$q_{air}^{HJ} = K' *(V_1 - V_e)^3$	Model data (W = 0.43 and 0.15 m). $1 < V_1 < 2.8$ m/s.
ERVINE and AHMED (1982)	Two-dimensional vertical dropshaft	$q_{air}^{HJ} = 0.00045*(V_1 - V_e)^3$	Vertical jets. $3 < V_1 < 6$ m/s.
RABBEN et al. (1983)	Hydraulic jump in horizontal rectangular pipe	$0.03 * (Fr_1 - 1)^{0.76}$	Model data. Gate opening ratio : 0.25.

Note : V_e :   inception velocity (table 3-4)

V_1 upstream flow velocity

Fr_1 upstream Froude number

3.5 Design of chutes with nappe flow regime

For the design of stepped cascades, SCHOKLITSCH (1937), at the beginning at the 20-th century, recommended to design stepped chutes with horizontal steps such as :

$$\frac{h}{l} = \frac{h}{L_d + 3 * h} \tag{3-29a}$$

and combining with equation (3-9) :

$$\frac{h}{l} = \left(3 + 4.3 * \left(\frac{d_c}{h} \right)^{0.81} \right)^{-1} \tag{3-29b}$$

Other authors suggested various criterion. A review of design criterion is presented in table 3-6, including pooled-step chutes.

Recently, a number of dams have been built in South Africa with stepped spillways. From this experience, STEPHENSON (1991) suggested that the most suitable conditions for nappe flow situations are :

$$\tan\alpha = \frac{h}{l} < 0.20 \tag{3-30a}$$

and $$\frac{d_c}{h} < \frac{1}{3} \tag{3-30b}$$

Equations (3-30) satisfy equation (3-14) : i.e., the flow situation satisfying both equations (3-30a) and (3-30b) is a nappe flow regime with fully developed hydraulic jump (sub-regime NA1).

STEPHENSON's (1991) recommendations imply relatively large steps and flat slopes. This situation is not often practical; but it may apply to relatively flat spillways, natural streams, creeks, river training and storm waterways. For steep channels or small step heights, a skimming flow regime is more desirable and will achieve larger rate of energy dissipation (chapter 5).

Table 3-6 - Design criterion for channel with nappe flow

Reference (1)	Geometry (2)	Tailwater level (3)	Comments (4)
Flat steps			
BINNIE (1913)	$h/l < 1/3$ $h < 0.305$ m (=1 ft)	N/A	Horizontal steps at upstream end.
SCHOKLITSCH (1937)	$\frac{h}{l} = \frac{h}{L_d + 3*h}$	N/A	Horizontal steps.
STEPHENSON (1991)	$h/l < 0.20$ $d_c/h < 1/3$	N/A	Horizontal steps.
Pooled steps			
BINNIE (1913)	$h/l = 1/10$	$d_t = h$	Pooled steps.
SCHOKLITSCH (1937)	$h/l = 1/3$	$d_t > 2$ m	Pooled steps

Note : d_t : tailwater depth (i.e. pool height)

3.6 Example of application

Considering a 5-degree slope stream bed with 10 steps, the river flows as a nappe flow regime. The total discharge is 8 m^3/s and the stream width is 20 m. The step height and length are : h = 1 m and l = 11.43 respectively. Compute the hydraulic characteristics of the flow.

Calculations

At each step, the entire flow conditions can be computed using equations (3-1) to (3-28). Correlations of plunging jet entrainment and hydraulic jump entrainment can provide a first estimate of the amount of entrained air. The main results are summarised in table 3-7.

Assuming a rectangular channel cross-section, the ratio d_c/h equals 0.25. With that value, equations (3-10) to (3-12) provide the nappe characteristics at the impact with the pool of water beneath the nappe. Downstream of the nappe impact, the flow depth upstream of the hydraulic jump is deduced from equation (3-6). The upstream Froude number of the jump (i.e. $Fr_1 = q_w/\sqrt{g^* d_1^3}$) equals 4.4.

Table 3-7 -Application for h = 1 m, Q_w = 8 m^3/s, W = 20 m, 10 steps

Variable (1)	Value (2)	Unit (3)	Eq. (4)	Comments (5)
d_c/h	0.254			
$(d_c/h)_{char}$	2.05		(3-14)	Fully developed hydraulic jump.
d_b	0.181	m	(3-1)	Flow depth at the brink of a step.
d_i	0.09	m	(3-10)	Nappe thickness at the impact with the receiving pool.
V_i	4.45	m/s	(3-11)	Impact velocity of the nappe.
θ	61.9	degrees	(3-12)	Jet angle of the impinging nappe.
d_1	0.094	m	(3-6)	Flow depth upstream of hydraulic jump
Fr_1	4.44			Froude number at start of hydraulic jump.
d_2	0.546	m	(3-7)	Flow depth downstream of hydraulic jump.
d_p	0.404	m	(3-8)	Flow depth in pool beneath the nappe.
L_d	1.42	m	(3-9)	Length of drop.
L_r	2.21	m	(3-13)	Roller length of the hydraulic jump.
$\Delta H/H_{max}$	90%		(3-16)	Rate of energy dissipation.
Q_{air}^{nappe}/Q_w	0.59		(3-17)	Nappe ventilation at each step.
V_e	1.14	m/s		Inception velocity of plunging nappe (KOGA 1982).
Fr	3.53			Nappe Froude number at the impact.
Q_{air}^{jet}/Q_w	0.080		(3-23)	Plunging jet entrainment at each step.
Q_{air}^{HJ}/Q_w	0.084		(3-26)	Air entrainment at hydraulic jump at each step.

Discussion

Equation (3-14) indicates that the flow is a nappe flow regime with fully-developed hydraulic jumps (sub-regime NA1). The knowledge of the nappe characteristics and flow depths at any point along the chute enables an accurate design of the chute sidewalls.

The rate of energy dissipation on the spillway is 0.90 (eq. (3-16)). At each step, an appropriate ventilation would be $q_{air}^{nappe} = 0.24 \ m^2/s$ (eq. (3-17)).

CHAPTER 4
SKIMMING FLOW REGIME (1)
HYDRAULICS OF SKIMMING FLOWS

4.1 Introduction

For large discharges, the waters flow down a stepped channel as a coherent stream, "skimming" over the steps. In the skimming flow regime, the external edges of the steps form a pseudo-bottom over which the water skims. Beneath the pseudo-bottom, recirculating vortices develop, filling the zone between the main flow and the steps. The vortices are maintained through the transmission of shear stress from the fluid flowing past the edges of the steps (fig. 4-1). In addition small-scale vorticity is generated continuously at the corner of the step bottom. Most of the flow energy is dissipated to maintain the circulation of the recirculation vortices.

On stepped chutes with skimming flow regime, the flow is highly turbulent and the conditions for free-surface aeration (eq. (1-1) and (1-2)) are satisfied. Large quantities of air are entrained along the channel (fig. 4-2). The free-surface aerated flow region follows a region where the flow over the chute is smooth and glassy. Next to the boundary, however, turbulence is generated and the boundary layer grows until the outer edge of the boundary layer reaches the surface (fig. 4-1). When the outer edge of the boundary layer reaches the free surface, the turbulence initiates natural free surface aeration. The location of the start of air entrainment is called the point of inception. Downstream of the inception point, a layer containing a mixture of both air and water extends gradually through the fluid. Far downstream the flow will become uniform and for a given discharge any measure of flow depth, air concentration and velocity distributions will not vary along the chute. This region is defined as the uniform equilibrium flow region.

The hydraulic characteristics of skimming flows are addressed in this section. First, the onset conditions for skimming flows are discussed. Then the calculations of the boundary layer growth are developed. Later the uniform flow properties are detailed taking into account the effects of air entrainment. The flow resistance calculations are developed and model data are presented. The effects of air entrainment are studied using results obtained with free-surface aerated flows on smooth chutes and on rockfilled channels. Although the flow conditions are different between un-stepped and stepped chutes, it is believed that the mechanisms of air entrainment are similar. Once the skimming flow becomes fully developed, a stepped chute behaves in the same way as an un-stepped one.

Fig. 4-1 - Flow regions above a stepped chute with skimming flow regime

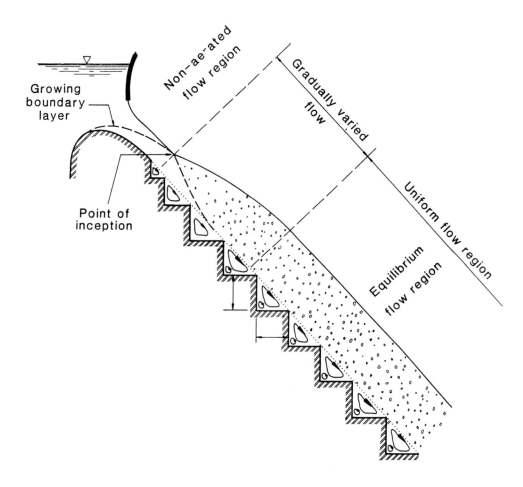

4.2 Onset of skimming flow

For small discharges and flat slopes, the water flows as a succession of waterfalls (i.e. nappe flow regime). An increase of discharge or of slope might induce the apparition of skimming flow regime. The "*onset of skimming flow*" is defined by the disappearance of the cavity beneath the free-falling nappes, the waters flowing as a quasi-homogeneous stream. The phenomenon presents some similarity with the cavity filling (or submergence) of aeration devices and ventilated cavities.

The onset of skimming flow is a function of the discharge, the step height and length. The author

re-analysed experimental data summarised in table 4-1. For the data, skimming flow regime
occurs for discharges larger than a critical value defined as :

$$\frac{(d_c)_{onset}}{h} = 1.057 - 0.465 * \frac{h}{l} \qquad (4-1)$$

where h is the step height, l is the step length and $(d_c)_{onset}$ is the characteristic critical depth.
Skimming flow regime occurs for $d_c > (d_c)_{onset}$, where d_c is the critical flow depth. Figure 4-3
compares equation (4-1) with experimental data. Equations (3-14) and (3-30) are shown also.
It must be emphasised that equation (4-1) was deduced for h/l ranging from 0.2 to 1.25 (i.e. 11 <
α < 52 degrees). There is no information on its validity outside of that range.

Fig. 4-2 - Air entrainment with skimming flow
Stepped spillway model of the Hydraulics Laboratory of CEDEX used by DIEZ-CASCON et al.
(1991) (Courtesy of Mr BANDLER) - α = 53.1 degrees, H_{dam} = 3.8 m, W = 0.8 m
(A) General view of the model

Fig. 4-2 - Air entrainment with skimming flow

Stepped spillway model of the Hydraulics Laboratory of CEDEX used by DIEZ-CASCON et al.
(1991) (Courtesy of Mr BANDLER) - α = 53.1 degrees, H_{dam} = 3.8 m, W = 0.8 m

(B) Details of the flow and the large amount of free-surface aeration

Table 4-1- Onset of skimming flow on stepped chutes- Experimental data

h/l	d_c/h	Reference
(1)	(2)	(3)
0.20	1.15	ESSERY and HORNER (1978)
0.42	0.81	
0.53	0.82	
0.74	0.82	
0.84	0.80	
0.33	0.74	PEYRAS et al. (1991),
0.5	0.67	DEGOUTTE et al. (1992)
1.0	0.61	
1.25	0.40	BEITZ and LAWLESS (1992)

Fig. 4-3 - Onset of skimming flow regime - Comparison between model data (ESSERY and
HORNER 1978, DEGOUTTE et al. 1992, BEITZ and LAWLESS 1992) and equation (4-1)

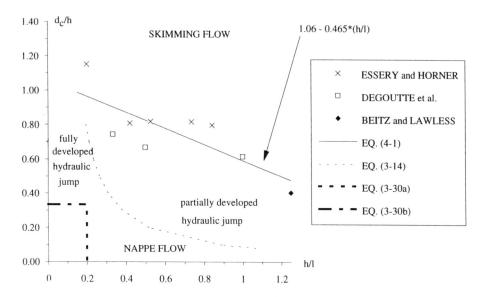

Discussion

The onset of skimming flow is similar to the onset of cavity filling of aeration devices and
ventilated cavities, recently reviewed by the author (CHANSON 1992c). When the local cavity
pressure falls below some critical value, the ventilated cavity can be drowned out and disappear.
For an aeration device with a 30-mm offset, on a 52-degree chute, the cavity filling occurred for
$d_b/h > 1.67$, d_b being the brink depth of supercritical flows (i.e. $4 < Fr_b < 15$).

For subcritical flows on flat channels with strip bottom roughness, several researchers observed
stable recirculatory motion in the grooves between adjacent elements. For all experiments (table
4-3), the occurrence of 'quasi-smooth flow' (also called 'skimming flow') was independent of the
flow rate and a function only of the roughness-spacing to roughness-height ratio :
$L_s/k_s \leq 3.4$ to 5 (table 4-2). For another experiment with cubical roughness elements
(O'LOUGHLIN and MACDONALD 1964), a stable recirculatory flow pattern was observed for
roughness densities larger than 0.25. For two-dimensional strip roughness, this condition would
become : $L_s/k_s \leq 4$. Note the close agreement between all the observations (table 4-2).

In horizontal channels and pipes, the onset of 'quasi-smooth flow' over artificial roughness was
defined to the apparition of stable recirculation vortices. It will be shown (paragraph 4.5) that the
definition of 'quasi-smooth flow' used by these researchers applies only to one particular type of
skimming flow : i.e., the "recirculating cavity flow" regime [SK3] characterised by stable

recirculating vortices. On stepped channels, skimming flows may occur with stable and unstable recirculation eddies in the cavities. Hence it is not possible to compare the apparition of stable recirculation vortices (table 4-2) with the onset of skimming flow and the disappearance of nappe flow (table 4-1).

Table 4-2 - Quasi-smooth flows over artificial roughness

Reference	Condition of 'quasi-smooth flow'	Remarks
(1)	(2)	(3)
ADACHI (1964)	$L_s/k_s \leq 3.4$	Two-dimensional strip roughness in open channel. Rectangular wooden bars.
O'LOUGHLIN and MACDONALD (1964)	Roughness density > 0.25	Cubical roughness elements in open channel. Plates cast of aluminium .
LEVIN (1968)	$\dfrac{\text{Groove length}}{\text{Groove depth}} < 4$ [a]	Pipe flow. Grooves perpendicular to the flow.
KNIGHT and MACDONALD (1979)	$L_s/k_s \leq 3.5$	Two-dimensional strip roughness in open channel. Square strips.
AKAMOTO et al. (1993)	$L_s/k_s \leq 5$	Two-dimensional strip roughness in wind and water tunnels. Square strips.

Note : [a] :   geometry of groove characterised by no increase in head loss

Fig. 4-4 - Geometry of strip roughness and large roughness

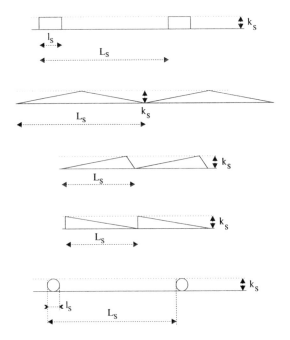

Table 4-3 - Experiments of flows over large roughness

Reference	Slope (deg.)	Discharge q_w (m^2/s)	Relative roughness k_s/D_H	Re	Remarks
(1)	(2)	(3)	(4)	(5)	(6)
Flows over strip roughness (water flow)					
BAZIN (1865) [a]	0.086 to 0.51	0.046 to 0.62	0.0077 to 0.055	1.6E+5 to 1.9E+6	Wooden rectangular strips : k_s = 10 mm, l_s = 27 mm, L_s = 37 mm. W = 1.98 m. Channel length > 200 m.
SKOGLUND (1936) [b]			0.008 to 0.06	4E+3 to 2.5E+5	Wide rectangular brass pipes. 60-degree V-grooves : k_s = 0.32 mm, L_s = 1.52 mm
STREETER (1936) [b]			0.006		Circular brass pipe ($\varnothing$ = 50.8 mm). Roughness type V : k_s = 3 mm, L_s = 11 mm.
			0.011		Roughness type VII : k_s = 5.6 mm, L_s = 22 mm.
STREETER and CHU (1949) [b] [c]			0.0112 & 0.0204	2E+5 to 8E+5	Circular aluminium pipe ($\varnothing$ = 0.114 m) with square threads.
ADACHI (1964)	0.11	7.5E-4 to 0.075	0.023 to 0.48		Wooden rectangular bars : k_s = 5 mm, l_s = 6.4 mm, W = 0.20 m.
TOWNES and SABERSKY (1966)		0.01 to 0.071	0.047 to 0.038	1E+4 to 7.1E+4	Square grooves : k_s = 3.175 to 25.4 mm, l_s/k_s = 0.5, L_s/k_s = 1.5, W = 0.85 m.
KNIGHT and MACDONALD (1979)	0.055	0.008 to 0.19	0.006 to 0.021	2.7E+4 to 3.2E+5	Perspex strip roughness of square cross-section : k_s = 3 mm, l_s = 3 mm, W = 0.46 m.
TOMINAGA and NEZU (1991)		0.0213	0.0378	1.4E+4	Two-dimensional circular strip roughness : $k_s(\varnothing_s)$ = 8 mm, L_s/k_s = 1 to 16, W = 0.4 m.
Flows over strip roughness (air flow)					
SAMS (1952) [b] [d] [e]			0.0084, 0.013 & 0.019	2E+4 to 3E+5	Circular Inconel pipe ($\varnothing$ = 12.7 mm) with square threads. Air flow.
PERRY et al. (1969) [e]					Recirculating wind tunnel. k_s = 3.2, 12.7 and 25.4 mm. l_s = 25.1, 12.7, & 22.3 mm. L_s/k_s = 1.83 & 9.
FURUYA et al. (1976) [e]		V = 13 and 21 m/s			Open circuit wind tunnel. Circular strip roughness : $k_s(\varnothing_s)$ = 2 mm, L_s/k_s = 1 to 64, W = 1 m.
OKAMOTO et al. (1993)		V = 12 and 16 m/s [e]			Wind and water tunnels. Square strip roughness : k_s = 10 mm, L_s/k_s = 2 to 17.
Flows over triangular roughness					
VITTAL et al. (1977)			0.05 to 0.13	3.2E+4 to 2.6E+5	Flat open channel with two-dimensional triangular roughness elements : [30 deg. triangle diagram] Roughness height : k_s = 0.03 m. Height-length ratio : 1/5. W = 0.6 m.
GEVORKYAN and KALANTAROVA (1992)			0.05 to 0.125		Stepped teeth directed against the flow: [diagram] Height-length ratio : 1/4. ZEMARIN's formula.

Table 4-3 - Experiments of flows over large roughness

Reference	Slope	Discharge	Relative roughness	Re	Remarks
	(deg.)	q_w (m^2/s)	k_s/D_H		
(1)	(2)	(3)	(4)	(5)	(6)
Flow over roughness elements					
SAYRE and ALBERTSON (1963)	0.06 to 0.17	0.02 to 0.07	0.05 to 0.14	8.3E+4 to 2.5E+5	Baffle blocks : k_s = 38 mm, L_s = 76 mm, width = 152 mm. W = 2.44 m.
O'LOUGHLIN and MACDONALD (1964)	0 to 2	up to 0.125	0.023 to 0.083		Cubical roughness elements : k_s = 12.7 mm. Spacing : 1.4 to 9*k_s. W = 0.61 m.

Notes : k_s, L_s, l_s roughness height, spacing and length (fig. 4-4)

       Re :   Reynolds number : Re = $\rho_w{}^*U_w{}^*D_H/\mu_w$

       [a] :   data reported also in POWELL (1944)

       [b] :   circular pipe flows

       [c] :   as reported in PERRY et al. (1969)

       [d] :   isotherm air flow

       [e] :   wind tunnel experiment

4.3 Boundary layer growth

4.3.1 Presentation

From the crest of the chute or from the upstream gate, a bottom turbulent boundary layer develops (fig. 4-1). The location where its outer edge reaches the free surface is called the inception point of air entrainment. Downstream of that location, the turbulence next to the free-surface become large enough to initiate natural free surface aeration (fig. 4-1). The characteristics at the inception point are L_I and d_I : L_I is the distance from the start of the growth of the boundary layer and d_I is the depth of flow at the point of inception.

On smooth spillways, the position of the point of inception is primarily a function of the discharge and the chute roughness. KELLER and RASTOGI (1977) suggested :

$$\frac{L_I}{k_s{'}} = f_1(F_*, \sin\alpha)$$
(4-2)

$$\frac{d_I}{k_s{'}} = f_2(F_*, \sin\alpha)$$
(4-3)

where F_* is a Froude number defined in terms of the roughness height : $F_* = q_w/\sqrt{g^*\sin\alpha^*(k_s{'})^3}$, where $k_s{'}$ is the skin roughness height and α is the channel slope.

CAIN and WOOD (1981) considered a boundary layer growth formula such as :

$$\frac{d_I}{L_I} = a_1 * \left(\frac{L_I}{k_s'}\right)^{-a_2} \tag{4-4}$$

in which a_1 and a_2 are constants. Assuming a velocity distribution of the form :

$$\frac{V}{V_I} = \left(\frac{y}{d_I}\right)^{1/N} \tag{4-5}$$

where V_I is the free surface velocity at the point of inception, and combining with the Bernoulli equation, equations (4-2) and (4-3) become (CAIN and WOOD 1981) :

$$\frac{L_I}{k_s'} = \left(\frac{N+1}{a_1 * N * \sqrt{2}}\right)^{2/(3-2*a_2)} * (F_*)^{2/(3-2*a_2)} \tag{4-6}$$

$$\frac{d_I}{k_s'} = (a_1)^{1/(3-2*a_2)} * \left(\frac{N+1}{N * \sqrt{2}}\right)^{(2-2*a_2)/(3-2*a_2)} * (F_*)^{(2-2*a_2)/(3-2*a_2)} \tag{4-7}$$

For smooth concrete spillways, WOOD et al. (1983) estimated the parameters a_1, a_2 and N as :

$$a_1 = 0.0212 * (\sin\alpha)^{0.11}$$

$$a_2 = 0.10$$

$$N = 6.0$$

Equations (4-6) and (4-7) become :

$$\frac{L_I}{k_s'} = 13.6 * (\sin\alpha)^{0.0796} * (F_*)^{0.713} \tag{4-8}$$

$$\frac{d_I}{k_s'} = \frac{0.223}{(\sin\alpha)^{0.04}} * (F_*)^{0.643} \tag{4-9}$$

and equation (4-4) yields :

$$\frac{d_I}{L_I} = 0.0212 * (\sin\alpha)^{0.11} * \left(\frac{L_I}{k_s'}\right)^{-0.10} \tag{4-10}$$

4.3.2 Application to stepped channels

On stepped chutes, the position of the start of air entrainment is a function of the flow discharge, crest design, bottom roughness, step geometry and chute geometry. Most designs of the ogee crest are fitted to a Creager profile (e.g. M'Bali dam) or a WES profile (e.g. Monksville dam). Usually, few smaller steps are introduced near the crest to eliminate deflecting jets of water (fig. 4-1). With such a complex geometry, the analysis of the growth of the boundary layer becomes extremely difficult.

The author re-analysed the flow properties at the point of inception of model experiments (table 1-1). For these data, the location of the start of air entrainment and the flow depth next to that location were recorded. The results are presented on figure 4-5 as L_I/k_s and d_I/k_s versus the Froude number F_*, where the roughness k_s was estimated as the depth of a step normal to the free surface (i.e. $k_s = h * \cos\alpha$) (fig. 4-6).

Fig. 4-5 - Characteristics of the inception point of air entrainment

(A) L_I/k_S as a function of F_* (BEITZ and LAWLESS 1992, BINDO et al. 1993, FRIZELL and MEFFORD 1991, SORENSEN 1985, TOZZI 1992)

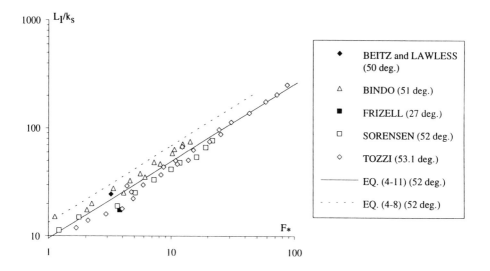

(B) d_I/k_S as a function of F_* (BINDO et al. 1993, FRIZELL and MEFFORD 1991, SORENSEN 1985, TOZZI 1992)

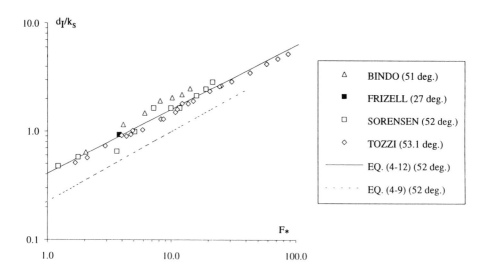

Fig. 4-6 - Definition of the roughness height k_s for a skimming flow regime

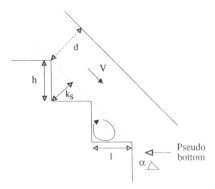

Fig. 4-7 - Effects of the discharge on the location of the inception of air entrainment (Courtesy of Mr ROYET, CEMAGREF) - Flow from the right to the left : α = 51.3 degrees, h = 0.06 m

(A) q_w = 0.057 m²/s

Fig. 4-7 - Effects of the discharge on the location of the inception of air entrainment (Courtesy of Mr ROYET, CEMAGREF) - Flow from the right to the left : α = 51.3 degrees, h = 0.06 m

(B) q_w = 0.079 m^2/s

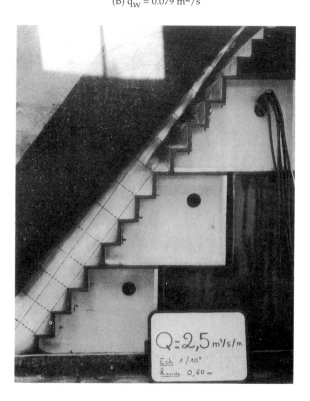

A statistical analysis of the data indicates that the flow properties are best correlated by :

$$\frac{L_I}{k_s} = 9.719 * (\sin\alpha)^{0.0796} * (F_*)^{0.713} \qquad (4\text{-}11)$$

$$\frac{d_I}{k_s} = \frac{0.4034}{(\sin\alpha)^{0.04}} * (F_*)^{0.592} \qquad (4\text{-}12)$$

where $F_* = q_w/\sqrt{g*\sin\alpha*(h*\cos\alpha)^3}$ and $k_s = h*\cos\alpha$. Equation (4-4) yields :

$$\frac{d_I}{L_I} = 0.06106 * (\sin\alpha)^{0.133} * \left(\frac{L_I}{k_s}\right)^{-0.17} \qquad (4\text{-}13)$$

Equations (4-11) and (4-12) are plotted on figure 4-5 with the model data.

A comparison between equations (4-8) and (4-11) shows that the application of smooth spillway calculations (eq. (4-8)) to stepped chutes is inaccurate. Smooth channel calculations overestimate the location of the apparition of 'white waters' on stepped chutes (fig. 4-5(A)). Indeed the rate of boundary layer growth on stepped chutes (eq. (4-13)) is approximately 2.8 times larger than on smooth channels (eq. (4-10)).

For a given channel geometry, equation (4-11) implies that the location of the inception point moves downstream with increasing discharge.

On figure 4-7, several photographs show a stepped spillway model (α = 51.3 deg., h = 0.06 m) with discharges between 0.057 and 0.266 m^2/s. For the smallest discharge, the point of inception is located near the top of the channel. For the largest discharge, the water is transparent all along the channel, indicating the absence of free-surface aeration : the outer edge of the turbulent boundary layer has not reached the free-surface before the downstream end of the channel.

Fig. 4-7 - Effects of the discharge on the location of the inception of air entrainment (Courtesy of Mr ROYET, CEMAGREF) - Flow from the right to the left : α = 51.3 degrees, h = 0.06 m

(C) q_w = 0.158 m^2/s

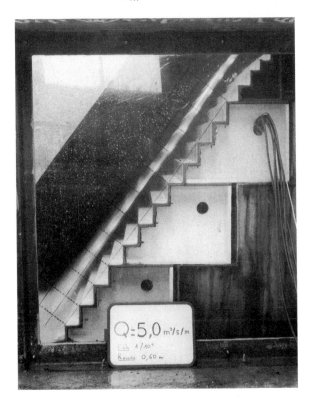

Fig. 4-7 - Effects of the discharge on the location of the inception of air entrainment (Courtesy of
Mr ROYET, CEMAGREF) - Flow from the right to the left : α = 51.3 degrees, h = 0.06 m

(D) $q_w = 0.266 \ m^2/s$

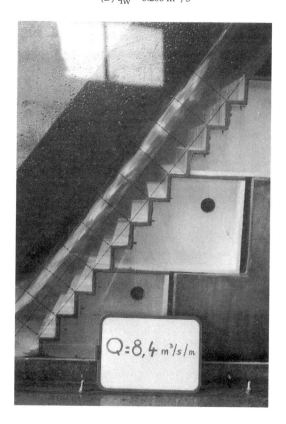

Discussion

Most ogee crests of the models were fitted to a smooth profile : e.g. a WES profile (SORENSEN
1985, TOZZI 1992), a Creager profile (BINDO et al. 1993) and a smooth crest close a WES profile
(BEITZ and LAWLESS 1992). But the model used by FRIZELL and MEFFORD (1991) did not
have a smooth intake : the crest was flat, followed directly by the first step. In any case, figure 4-5
suggests that the type of ogee crest profile might have little influence on the characteristics of the
inception point.

It must be emphasised that equations (4-11) and (4-12) were deduced from data obtained with
channel slopes ranging from 27 to 53 degrees. Great care must be taken when applying these
equations outside of that range.

4.4 Uniform flow conditions

If the chute is long enough, uniform flow conditions are reached before the end of the chute (fig. 4-1). In this paragraph, the uniform flow characteristics are analysed using the continuity and momentum equations.

Definitions

On stepped channels, visual observations indicate a large amount of air being entrained through the flow free-surface (fig. 4-2). And it is necessary to define appropriate air-water flow parameters. The following definitions are used commonly in free-surface aerated flows.

The local air concentration C is defined as the volume of air per unit volume of air and water. The characteristic water flow depth d is defined as :

$$d \ = \ \int_0^{Y_{90}} (1 - C) * dy \qquad\qquad (4\text{-}14)$$

where y is measured perpendicular to the channel surface and Y_{90} is the depth where the local air concentration is 90%. The characteristic depth d is defined from 0 to Y_{90} : above 90% of air concentration, air concentration and velocity measurements are not accurate (CHANSON 1992a) and the integration of the air concentration above Y_{90} becomes meaningless.

The depth averaged mean air concentration is defined as:

$$C_{mean} \ = \ \frac{1}{Y_{90}} * \int_0^{Y_{90}} C * dy \qquad\qquad (4\text{-}15a)$$

and combining with equation (4-14), this yields :

$$C_{mean} \ = \ 1 \ - \ \frac{d}{Y_{90}} \qquad\qquad (4\text{-}15b)$$

The mean flow velocity is defined as :

$$U_w \ = \ \frac{q_w}{d} \qquad\qquad (4\text{-}16)$$

where q_w is the water discharge per unit width. The characteristic velocity V_{90} is defined as that at Y_{90}.

4.4.1 Flow aeration

The amount of air entrained within the flow is defined usually in term of the depth averaged mean air concentration[1]. The analysis of free-surface aerated flow measurements on smooth chutes (table 4-4) showed that the average air concentration for uniform flows C_e is independent

---

[1]The quantity of air entrained within the flow is related to the depth averaged mean air concentration by :
$q_{air}/q_w = C_{mean}/(1 - C_{mean})$.

of the upstream geometry and flow conditions (i.e. discharge, flow depth, roughness) and is a function of the slope only (WOOD 1983, CHANSON 1993a).

On figure 4-8, the mean equilibrium air concentration C_e is plotted as a function of the channel slope α for model data (STRAUB and ANDERSON 1958) and field data (AIVAZYAN 1986). Table 4-5 summarises the results presented on figure 4-8. For slopes flatter than 50 degrees, the average air concentration may be estimated as :

$$C_e = 0.9 * \sin\alpha \tag{4-17}$$

In mountain rivers and rockfill channels, visual and photographic observations show clearly a large amount of free-surface aeration (i.e. 'white waters') as illustrated by JUDD and PETERSON (1969), BATHURST (1978) and THORNE and ZEVENBERGEN (1985). KNAUSS (1979) analysed the data of HARTUNG and SCHEUERLEIN (1970) obtained with great natural roughness and steep slopes (table 4-6). He indicated that the quantity of air entrained on rockfill channels was estimated as :

$$C_e = 1.44 * \sin\alpha - 0.08 \tag{4-18}$$

This result is of similar form as equation (4-17).

Figure 4-8 compares both equations (4-17) and (4-18) with experimental data obtained on smooth chutes. The results show a comparable rate of air entrainment for both smooth and rough channels. It is expected that the uniform air concentration for stepped chute flows is similar to these results, for which the mean equilibrium air concentration is a function of the slope only (table 4-5) (CHANSON 1993c).

Fig. 4-8 - Uniform equilibrium air concentration C_e as a function of the chute slope α - Model data (STRAUB and ANDERSON 1958), prototype data (AIVAZYAN 1986) and rockfill channel (KNAUSS 1979, eq. (4-18))

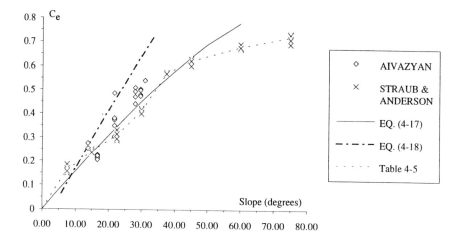

Table 4-4 - Smooth chute experimental data

Chute/Spillway	Slope degrees	q_w m²/s	k_s'/D_H	Re	Ref.	Comments
(1)	(2)	(3)	(5)	(6)	(7)	(8)
Prototypes						
Ak-Tepe, Russia (former USSR)	21.8	2.3 to 8.0	5E-3 to 1E-2	8.8E+6 to 2.8E+7	[A]	Rough concrete. W = 5 m. k_s' = 5 mm.
Aviemore, New Zealand	45.0	2.23 & 3.16	8E-4 to 1.9E-3	8.9E+6 to 1.3E+7	[CA]	Concrete. k_s' = 1 mm.
Bencok, Indonesia	31.05	2.9 to 6.0	3E-3 to 4.6E-3	9E+6 to 1.6E+7	[A,E,L]	W = 1 m. k_s' = 2 mm.
Big Hill, Australia	4.2	0.74 to 0.82	6.5E-3	2E+6	[M]	Smooth concrete.
Boise, USA	4.6 to 12.2	0.01 to 0.8	2.7E-3 to 3.1E-2	5.3E+4 to 2.8E+6	[S]	Concrete. W = 0.9 to 1.8 m. k_s' = 1 mm.
Dago, Indonesia	13.8	0.74 to 1.39	2.5E-3 to 3.4E-3	2.5E+6 to 4.5E+6	[A,E,L]	W = 1 m. k_s' = 1 mm.
Erevan, Armenia (former USSR)	21.8	0.38 to 1.55	1.9E-2 to 4E-2	1.5E+6 to 5.8E+6	[A]	Rough basalt and cement mortar. W = 4 m. k_s' = 10 mm.
Gizel'don, Armenia (former USSR)	28.1	0.49 to 1.28	5E-4 to 1.1E-3	1.9E+6 to 5E+6	[A]	Wooden flume. W = 6 m. k_s' = 0.3 mm.
Hat Creek, USA	23.45 to 34.75	1.86 to 6.4	4.8E-3 to 1.2E-2	6E+6 to 2E+7	[H]	Rough concrete. W = 1.75 m. k_s' = 5 mm.
Kittitas, USA	33.2	2.24 to 11.7	7E-3 to 2E-2	8E+6 to 3E+7	[H]	Eroded concrete. W = 2.44 m. k_s' = 10 mm.
Mallnitz, Austria	22.2				[A,E,L]	Concrete. W = 2m.
Mostarsko Blato, Bosnia-Hercegovina (former Yugoslavia)		0.71 to 3.44	1.5E-2 to 3.5E-2	8.3E+4 to 3E+7	[J]	Stone lining. W = 5.35 m. k_s' = 20 mm.
Rapid Flume, USA	20.1	1.76	3E-4	6E+6	[H]	Wooden flume. W = 1.4 m. k_s' = 0.1 mm.
Spring Gully, Australia	5.3		6E-3 to 7E-4	2E+6	[M]	Smooth concrete. Semi circular (R=0.61 m).
Rutz, Austria	34	0.4 to 2	3E-4 to 1.3E-3	2E+6 to 7E+6	[I]	Concrete. Trapezoidal.
Spillway models						
Clyde model, New Zealand	52.3	0.21 to 0.48	3.8E-4 to 1.2E-3	8E+5 to 2E+6	[CH]	Perspex. D/S of aerator. W=0.25 m. k_s' = 0.1 mm.
IWP, Armenia (former USSR)	16.7 & 29.7	0.064 to 0.13	1.3E-3 to 2.6E-3	2.3E+5 to 4.3E+5	[A]	Planed board with painting. W=0.25 m. k_s' = 0.1 mm.
Vienna Lab., Austria	8.7 to 31.2	0.04 to 0.178			[A,E,J,L]	Wooden flume. W=0.25 m.
St Anthony Falls, USA	7.5 to 75	0.14 to 0.93	3E-3 to 1.6E-2	4.7E+5 to 2E+6	[SA]	Artificial roughness. W=0.46 m. k_s' = 0.7 mm.

Notes : k_s' surface (skin) roughness height

Re   Reynolds number : Re = $\rho_w * U_w * D_H / \mu_w$

[A] AIVAZYAN (1986); [CA] CAIN (1978); [CH] CHANSON (1988); [E] EHRENBERGER (1926);

[H] HALL (1943); [I] INNEREBNER (1924); [J] JEVDJEVICH and LEVIN (1953) [L] LEVIN (1955);

[M] MICHELS and LOVELY (1953); [S] STEWART (1913); [SA] STRAUB and ANDERSON (1958).

Table 4-5 Average air concentration in uniform self-aerated flows

Slope α	C_e [a]	Y_{90}/d_o	f_e/f	h/l
degrees	[b]	[b]	[c]	
(1)	(2)	(3)	(4)	(5)
0.0	0.0	1.0	1.0	0
7.5	0.1608	1.192	0.964	0.132
15.0	0.2411	1.318	0.867	0.268
22.5	0.3100	1.449	0.768	0.414
30.0	0.4104	1.696	0.632	0.577
37.5	0.5693	2.322	0.430	0.767
45.0	0.6222	2.647	0.360	1.0
60.0	0.6799	3.124	0.277	1.732
75.0	0.7209	3.583	0.215	3.732

Notes : ([a]) : as defined in equation (4-15); ([b]) : data from STRAUB and ANDERSON (1958);

      ([c]) : computed from equation (4-36)

Table 4-6 - Experimental data on rockfilled channels

Reference	Slope α	Discharge q_w	Relative roughness k_s/D_H	Re	Remarks
	(deg.)	(m²/s)			
(1)	(2)	(3)	(4)	(5)	(6)
JUDD and PETERSON (1969)	0.5 to 3.8	0.06 to 3	0.04 to 0.72 [a]	1.9E+5 to 8.8E+6	Field data. Natural torrents in USA.
HARTUNG and SCHEUERLEIN (1970)	6 to 34		0.02 to 0.2	8.5E+4 to 2E+6	Model study. Artificial rockfilled channel, Germany.
BATHURST (1978)	0.5 to 1	0.06 to 0.37	0.11 to 0.34 [a]	2.1E+5 to 1.4E+6	Field data. Natural streams in England.
THOMPSON and CAMPBELL (1979)	0.2 to 3	0.3 to 7.7	0.014 to 0.15 [a]	1.2E+6 to 2.8E+7	Field data. Torrents in USA and rockfilled channels in New Zealand.
BATHURST (1985)	0.23 to 2.2	0.02 to 4.9	0.012 to 0.25 [a]	9.5E+4 to 1.7E+7	Field data in United Kingdom.
THORNE and ZEVENBERGEN (1985)	0.8 to 1.1	0.15 to 0.9	0.065 to 0.11	6E+5 to 3.5E+6 [a]	Field data. Mountain river in Colorado, USA.
BATHURST (1988)	2.1 to 3.4	0.09 to 1	0.045 to 0.1	1.2E+6 to 4.3E+6	Field data in Colorado, USA. W = 5 to 6.2 m.
GOMEZ (1993)	0.5 to 2.3	0.09 to 0.17	0.028 to 0.051 [a]	2.5E+5 to 4E+5	Model study (W = 0.5 m).
FERRO and BAIAMONTE (1994)		0.01 to 0.16	0.035 to 0.14 [a]	2.6E+4 to 4.1E+5	Model study (W = 0.6 m). Series IV.

Notes : Re :   Reynolds number : $Re = \rho_w {}^* U_w {}^* D_H / \mu_w$

    ([a]) : k_s is the median boulder diameter (d_{50})

4.4.2 Flow properties

Assuming a long stepped channel and if the uniform flow conditions are reached before the end of the channel, the uniform flow depth is deduced from the momentum equation : i.e., the weight component in the flow direction equals the bottom friction. It yields :

$$\tau_o * P_w = \rho_w * g * A_w * \sin\alpha \tag{4-19}$$

where P_w is the wetted perimeter, ρ_w is the water density, g is the gravity constant, A_w is the water flow cross-section area. τ_o is the average shear stress between the skimming flow and the recirculating fluid underneath : i.e., τ_o is the mean shear stress along the pseudo-bottom. By analogy with clear water flows, the average shear stress τ_o is defined as for an open channel flow (HENDERSON 1966, STREETER and WYLIE 1981) :

$$\tau_o = \frac{f_e}{8} * \rho_w * (U_w)_o{}^2 \tag{4-20}$$

where f_e is the Darcy friction factor of the air-water flow and $(U_w)_o$ is the uniform velocity of the flow. Combining equations (4-19) and (4-20), it yields :

$$(U_w)_o = \sqrt{\frac{8 * g}{f_e}} * \sqrt{\frac{D_H}{4}} * \sin\alpha \tag{4-21}$$

where D_H is the hydraulic diameter : $D_H = 4 * A_w / P_w$.

For a wide channel, the uniform flow velocity $(U_w)_o$ and normal depth d_o are deduced from the continuity and momentum equations. In dimensionless form, it yields :

$$\frac{(U_w)_o}{V_c} = \sqrt[3]{\frac{8 * \sin\alpha}{f_e}} \tag{4-22}$$

$$\frac{d_o}{d_c} = \sqrt[3]{\frac{f_e}{8 * \sin\alpha}} \tag{4-23}$$

where V_c is the critical flow velocity. Combining equations (4-15) and (4-23), the characteristic depth $(Y_{90})_o$ for uniform flow becomes :

$$\frac{(Y_{90})_o}{d_c} = \sqrt[3]{\frac{f_e}{8 * (1 - C_e)^3 * \sin\alpha}} \tag{4-24}$$

where C_e is deduced from table 4-5. The depth $(Y_{90})_o$ takes into account the flow bulking caused by the air entrainment and it may be used as a design parameter for the height of the chute sidewalls.

For a given channel slope, figure 4-8 and equation (4-17) provide an estimate of the mean air concentration for uniform flow. Equations (4-21) to (4-24) enable the calculations of the uniform flow parameters as functions of the water discharge and friction factor. The calculations of the friction factor f_e are developed in the next paragraph.

4.5 Flow resistance

4.5.1 Presentation

Several authors studied skimming flows past large roughness elements in pipes and open channels (table 4-3). Their results indicated that the classical flow resistance calculations must be modified to take into account the shape of the roughness elements. Indeed the flow resistance is the sum of the skin resistance and the form resistance of the steps. For a stepped chute, the geometry of the step is characterised by its depth normal to the streamlines (i.e. $k_s = h*\cos\alpha$) and the channel slope (i.e. $\tan\alpha = h/l$) (fig. 4-6).

Dimensional analysis suggests that the friction factor is a function of the surface (skin) roughness height k_s', the Reynolds number, the (step) roughness height k_s, the channel slope and the quantity of air entrained :

$$f_e = f_3\left(\frac{k_s'}{D_H}; Re; \frac{k_s}{D_H}; \alpha; C_{mean}\right) \tag{4-25}$$

where the Reynolds number is defined as : $Re = \rho_w*U_w*D_H/\mu_w$, μ_w is the dynamic viscosity of water and C_{mean} is the average air concentration. In an uniform flow situation, C_{mean} and U_w equal respectively C_e and $(U_w)_o$.

Note

Some engineers compute the flow resistance using the Manning formula. For clear water flows, the Manning coefficient $n_{Manning}$ and the friction factor f are related by

$$n_{Manning} = \sqrt{\frac{f}{8*g}} * \left(\frac{D_H}{4}\right)^{1/6}$$

The use of the Manning formula implies that the velocity distributions follow a 1/6-th power law (CHEN 1990). Such an assumption is not realistic for extremely rough flows and for stepped channel flows (see paragraph 4.6). The Manning coefficient must be avoided for hydraulic calculations of stepped channels.

4.5.2 Analysis of experimental data

Several investigators (table 1-1) measured flow depths along and at the bottom of stepped spillway models and chutes. The re-analysis of the data provides new information on the flow resistance. It must be noted that all these measurements neglected the effects of flow aeration.

If the uniform flow conditions are reached along a constant slope channel, the Darcy friction factor is deduced from the momentum equation (4-19). Neglecting the flow aeration, it yields :

$$f = \frac{8*g*d_o^2}{q_w^2} * \frac{D_H}{4} * \sin\alpha \tag{4-26}$$

where f is the non-aerated flow friction factor, q_w is the water discharge per unit width. For non-uniform gradually varied flows, the friction factor can be deduced from the energy equation :

$$f = \frac{8 * g * d^2}{q_w^2} * \frac{D_H}{4} * \frac{\Delta H}{\Delta s} \qquad (4\text{-}27)$$

where ΔH is the total head loss over a distance Δs. $\Delta H/\Delta s$ is the friction slope (HENDERSON 1966).

The re-analysis of model data using equations (4-26) and (4-27), and neglecting the flow aeration, provides a first estimate of the non-aerated friction factor f as a function of the Reynolds number, relative roughness and channel slope :

$$f = f_4\left(\frac{k_s'}{D_H} ; Re ; \frac{k_s}{D_H} ; \alpha\right) \qquad (4\text{-}28)$$

A detailed analysis of the data indicates that the friction factor is independent of the surface roughness k_s' and of the Reynolds number Re. And equation (4-28) becomes :

$$f = f_5\left(\frac{k_s}{D_H} ; \alpha\right) \qquad (4\text{-}29)$$

The results are presented in figures 4-9 and 4-10 where the non-aerated friction factor f is plotted as a function of the relative (step) roughness k_s/D_H. The information on the channel slope is reported in the legend and in table 1-1.

Figure 4-9 presents the data obtained with flat chutes (i.e. $\alpha < 12$ degrees) while experimental data obtained with steep slopes (i.e. $\alpha > 25$ degrees) are presented on figure 4-10.

For flat channels, the results show a good agreement between the model data (NOORI 1984) and the prototype data (GRINCHUK et al. 1977). Further, figure 4-9 indicates a reasonable correlation between the friction factor and the relative roughness : i.e., an increase of the friction factor with k_s/D_H that appears to be independent of the channel slope. And equation (4-29) can be correlated by :

$$\frac{1}{\sqrt{f}} = 1.42 * Ln\left(\frac{D_H}{k_s}\right) - 1.25 \qquad \text{(Flat slopes, sub-regime SK1)} \quad (4\text{-}30)$$

Equation (4-30) was obtained for $0.02 < k_s/D_H < 0.3$. It should not be used outside of that range.

For steep slopes, the experimental results (fig. 4-10) show little correlation between the flow resistance, the relative roughness and the channel slope. For channel slopes between 50 and 55 degrees, figure 4-10 indicates values of the friction factor in the range 0.17 to 5 with a mean value of about 1.0. An earlier publication (CHANSON 1993c) suggested to use a mean value of $f = 1.3$ for skimming flows. The present investigation is more complete and the author recommends to use $f = 1.0$ as an *order of magnitude* of the friction factor for skimming flow on steep slopes (sub-regime SK3).

Note again that the experimental data were analysed neglecting the effects of air entrainment. No information is available on the amount of air entrained during the experiments. The author believes that the flow depth measurements overestimate the clear water flow depth. As a result the values of the friction factor shown on figures 4-9 and 4-10 could be overestimated.

Fig. 4-9 - Friction factors of stepped channels - Flat slopes (i.e. $\alpha < 12$ degrees) (table 1-1)

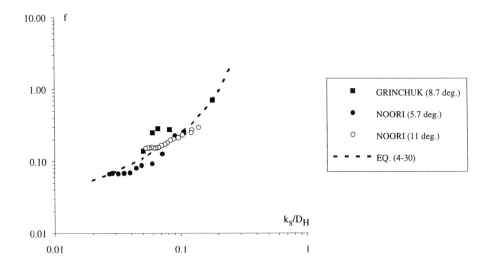

Fig. 4-10 - Friction factors of stepped channels - Steep slopes (i.e. $\alpha > 27$ degrees) (table 1-1)

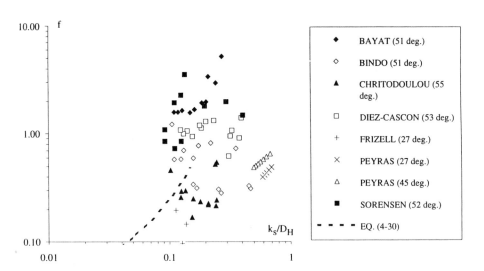

4.5.3 Discussion

A comparison between figure 4-9 and figure 4-10 indicates a substantial increase of the friction resistance from {f ~ 0.1} for flat stepped channels (α ~ 10 degrees, fig. 4-9) up to {f ~ 1} on steep stepped chutes (α ~ 50 degrees, fig. 4-10). Such a difference reflects different flow patterns in the recirculating cavities beneath the skimming stream.

For flat channels (α ~ 10 degrees), the cavity of recirculating fluid beneath the pseudo-bottom formed by the step edges is oblong and relatively thin, and large stable recirculating eddies cannot develop (fig. 4-11(A)). The recirculation vortices do not fill the entire cavity between the edges of adjacent steps. The wake from one edge interferes with the next step as shown by photographs (fig. 4-12). The vortex generation and dissipation process associated with each wake are disturbed by the next step and might interfere with those of the adjacent steps. The energy dissipation and flow resistance become functions of the wake development. The flow depth will control in part the vertical extent of the wake and the vortex interference region. The main flow parameters become the distance between two adjacent step edges and the depth of flow. This reasoning is confirmed by the results shown on figure 4-9 which shows a good correlation between the friction factor and the relative roughness, the latter being proportional to the edge spacing to flow depth ratio.

For very flat slopes, the flow is characterised by the impact of the wake on the next step, a three-dimensional unstable recirculation in the wake and some skin friction drag on the step downstream of the wake impact (fig. 4-11(A)). The flow pattern is called a "wake-step interference" sub-regime [SK1]. For larger slopes, the wake interferes with the next wake and there is no skin-friction drag component (fig. 4-11(B)). This pattern is called a "wake-wake interference" sub-regime [SK2].

For steep slopes, a stable recirculation in the cavities between the edges of adjacent steps is observed as shown by photographs (PEYRAS et al. 1991, DIEZ-CASCON et al. 1991). The recirculating eddies are large two-dimensional and possibly three-dimensional vortices (fig. 4-11(C)). The energy dissipation and the flow resistance are functions of the energy required to maintain the circulation of these large-scale vortices. The flow pattern is called a "recirculating cavity flow" sub-regime [SK3].

For flows past two-dimensional cavity, various researchers (table 4-7) observed a mechanism of stable recirculation (e.g. fig. 4-13) for a range of cavity dimensions. Photographic evidences of stable flow recirculation can be found in TOWNES and SABERSKY (1966) and FURUYA et al. (1976). Most researchers observed stable re-circulatory flow patterns for groove height to length ratio larger than 0.4 to 0.45. For a stepped channel, the cavity height to length ratio equals : [$\cos\alpha * \sin\alpha$] and this value of 0.4 would correspond to a channel slope α = 27 degrees. The results summarised in table 4-7 would imply that stable recirculation occurs for slopes larger than 27 degrees on a stepped chute. The comparison between figures 4-9 and 4-10 indicates a different flow resistance behaviour for channel slopes larger than 27 degrees, and this reasoning is coherent with the findings of other researchers (table 4-7).

Fig. 4-11 - Flow patterns in the cavity between adjacent steps

(A) Wake-step interference sub-regime [SK1] (flat slopes)

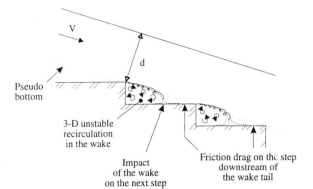

(B) Wake-wake interference sub-regime [SK2] (slope about 27 degrees)

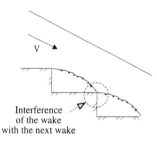

(C) Recirculating cavity flow sub-regime [SK3] (steep slopes)

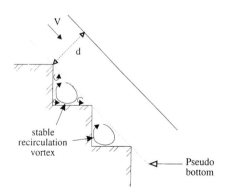

Fig. 4-12 - Skimming flow with wake-step interference sub-regime SK1 past overlapping concrete blocks (BAKER 1990) (Courtesy of Dr BAKER, University of Salford)

Flow from the right to the left ($q_W = 0.17 \, m^2/s$, W = 0.6 m)

α = 21.8 deg., h = 0.019 m, inclined downward steps : δ = - 8.3 deg.

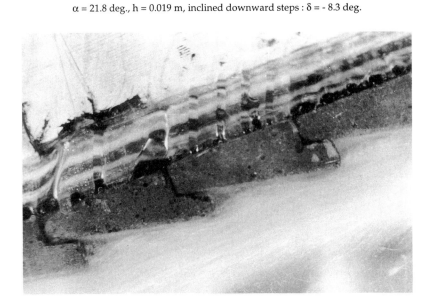

Table 4-7 - Observations of stable re-circulatory flow motion in two-dimensional cavity

Reference	Condition for stable recirculation	Remarks
(1)	(2)	(3)
Water flow experiments ADACHI (1964)	$\dfrac{L_s - l_s}{k_s} \leq 2.4$	Rectangular wooden bars.
TOWNES and SABERSKY (1966)	$\dfrac{V_* * k_s}{v} \geq 150$	Square grooves (i.e. $L_s - l_s = k_s$).
KNIGHT and MACDONALD (1979)	$\dfrac{L_s - l_s}{k_s} \leq 2.5$	Two-dimensional strip roughness in open channel. Square strips.
Wind tunnel experiments MAULL and EAST (1963)	$\dfrac{L_s - l_s}{k_s} \leq 2.5$	Flow past a two-dimensional rectangular cavity.
KISTLER and TAN (1967)	$\dfrac{L_s - l_s}{k_s} \leq 2.2$	Flow past a two-dimensional rectangular cavity.
FURUYA et al. (1976)	$\dfrac{L_s - l_s}{k_s} \leq 2$	Two-dimensional circular strip roughness. [a]

Notes : V_* shear velocity

($L_S - l_S$) groove or cavity length measured parallel to the flow direction (fig. 4-13)

[a]   ($L_S - l_S$) is measured between the outer edges of the circular rods (fig. 4-4)

Fig. 4-13 - Example of stable flow recirculation observed in a two-dimensional cavity

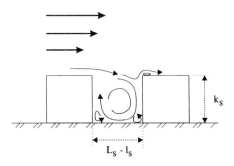

For slopes around 27 degrees, the flow pattern (sub-regime SK2) is characterised by the interference of the wake tail with the next wake (fig. 4-11(B)). Note the lack of data available for slopes ranging from 15 to 45 degrees.

4.5.4 Comparison with flows above large roughness

Flows past large roughnesses and flows above rockfilled channels can exhibit skimming flow regimes (fig. 4-14 and 4-15). A new comparative analysis between these skimming flow regimes and skimming flow situations above stepped channels is developed below. It provides some challenging results.

Fig. 4-14 - Skimming flow above large roughnesses

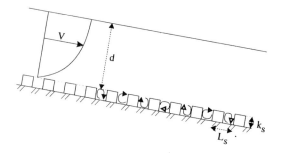

Fig. 4-15 - Skimming flow above a rockfilled channel

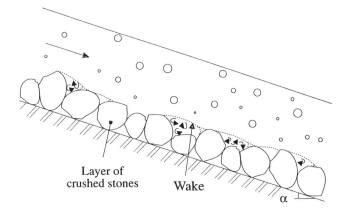

Layer of
crushed stones   **Wake**

α

4.5.4.1 Flow resistance of triangular roughness

VITTAL et al. (1977) performed experiments in open channel with two-dimensional triangular roughness (table 4-3). For a roughness height to length ratio of 1/5, their results were presented as :

$$f = 1.44 * \frac{k_s}{L_s} * \left(\frac{d}{k_s} + \frac{1}{2}\right)^{-3/8} \tag{4-31}$$

where L_s is the roughness spacing. The height-length ratio of 1/5 would correspond to a stepped channel slope of 11.3 degrees.

GEVORKYAN and KALANTAROVA (1992) considered the flow past stepped teeth opposed to the flow (table 4-3). For a roughness height to length ratio of 1/4, they reported the flow resistance formula of ZEMARIN which was obtained experimentally :

$$f = 8 * g * \left(\frac{1000}{52 - 5.1 * \frac{d}{k_s}}\right)^{-2} \tag{4-32}$$

Equations (4-31) and (4-32) are compared on figure 4-16. They are labelled VITTAL and GEVORKYAN respectively. The plot indicates a close agreement between the two equations.

4.5.4.2 Flow resistance of large roughness elements

Figure 4-16 presents a re-analysis of friction factor values for skimming flows above large roughness : cubical roughness, baffle blocks, two-dimensional strip roughness. The experimental data were obtained with open channel flows and pipe flows (table 4-3).

The results (fig. 4-16) indicate a reasonable agreement between the various experiments. The trend shown on figure 4-16 indicates an increase of the friction factor with the relative roughness and shows that f is independent of the Reynolds number.

Fig. 4-16 - Friction factors of flows above artificial roughness (table 4-3)

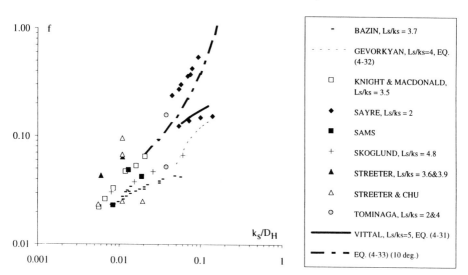

4.5.4.3 Flow resistance of rockfilled channels

The author re-analysed experimental data obtained in mountain rivers and rockfill channels using equation (4-26). The results are presented in figure 4-17. Details of the rockfill channel experiments are reported in table 4-6. Note that these data neglect the effect of air entrainment.

For HARTUNG and SCHEUERLEIN's (1970) experiments and neglecting air entrainment, SCHEUERLEIN (1973) summarised the results as :

$$\frac{1}{\sqrt{f}} = -3.2 * \text{Log}_{10}\left((1.7 + 8.1 * \sin\alpha) * \frac{k_S}{D_H}\right) \qquad (4\text{-}33)$$

Equation (4-33) is plotted on figure 4-17 and compared with the other experimental data. HARTUNG and SCHEUERLEIN (1970) choose arbitrarily k_S as one third of the equivalent diameter of the stones while other researchers used the median boulder diameter d_{50} to characterise the roughness height (table 4-6).

Figure 4-17 indicates an increase of the friction factor with the relative roughness. The analysis of the data shows no correlation with the Reynolds number. It is worth noting that both figure 4-16 and figure 4-17 exhibit the same trend : that is, an increase of the friction factor with the relative roughness k_S/D_H, where k_S is the height of the roughness elements.

Fig. 4-17 - Friction factors of flows above rockfilled channels (table 4-6)

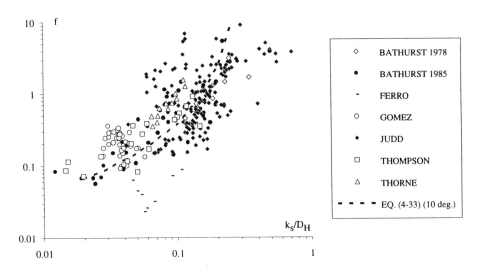

Fig. 4-18 - Comparison of the flow resistance on flat stepped channels, on rockfill channels and over large roughness

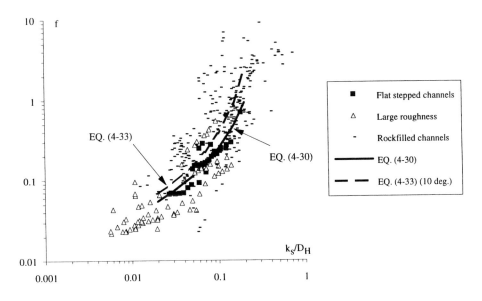

4.5.4.4 Discussion

Figure 4-18 compares experimental data obtained on stepped chutes with flat slopes (i.e. $\alpha < 12$ degrees), on rockfilled channels and over large roughness. All the sets of data indicate a similar trend : i.e., the friction factor increasing with the relative roughness k_s/D_H.

More figure 4-18 shows a good agreement between the experiments over large roughness, experiments on stepped channels and rockfilled channel data. It is thought that these flows have similar patterns and are dominated by unstable recirculating eddies and wake interference processes. Flows over rockfilled channels do not exhibit steady stable recirculating flow motion but unstable vortices behind rocks (fig. 4-15). These three-dimensional unstable flow patterns have some similarity with the flow patterns above flat stepped channels (sub-regime SK1, fig. 4-11(A)). This may explain the close values of the friction factors (fig. 4-18). Note that equations (4-30) and (4-33) provide close results.

On steep stepped channels (sub-regime SK3), the large-scale recirculating eddies play a major role in dissipating the flow energy. Hence, the flow resistance is directly related to the recirculation mechanisms rather than to the outer flow characteristics. This reasoning is illustrated by the apparent lack of correlation between the friction factor data, the Reynolds number and the relative roughness. Presently, no way of predicting the flow resistance is known. A similar conclusion was obtained by PERRY et al. (1969) who analysed velocity profiles in developing boundary layers in arbitrary pressure gradients.

HAUGEN and DHANAK (1966) and KISTLER and TAN (1967) measured turbulent shear stresses in shear layers between recirculating cavities and turbulent boundary layer flows. They observed that the maximum shear stress occurs along the pseudo-bottom formed by the cavity edges with typical values $\tau_o/(0.5*\rho*V^2)$ ranging from 0.005 up to 0.015.

RAJARATNAM (1990) developed a challenging comparison between experiments in stepped channels and experimental results obtained with steeppass and Denil fishways. His analysis showed that the friction factors for fishways are of the same order of magnitude as for stepped channels. For steeppass fishways, RAJARATNAM and KATOPODIS (1991) obtained friction factors in the range 0.4 to 4.

4.5.5 Effects of air entrainment on the flow resistance

The above analysis is based upon experimental data obtained with measurements neglecting the effects of flow aeration. Further the stepped channel data were obtained with small scale models. Several researchers (JEVDJEVICH and LEVIN 1953, WOOD 1983, CHANSON 1993a) showed that the presence of air within turbulent boundary layers reduces the shear stress between the flow layers and hence the shear force. The resulting drag reduction induces a reduction of the energy dissipation on the chute : as a result, the efficiency of a stepped channel, in term of energy dissipation, is diminished.

Fig. 4-19 - Drag reduction caused by free-surface aeration observed on smooth chutes
Comparison between model and prototype data (table 4-4) and equation (4-36)

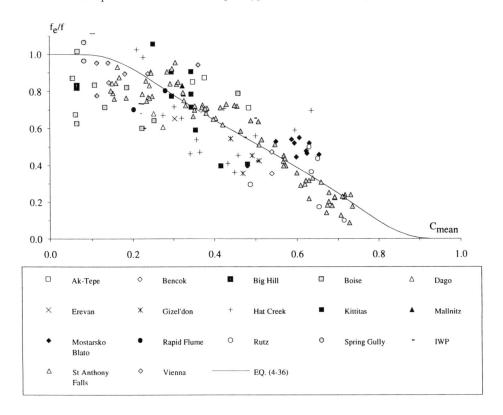

The effect of air entrainment on skimming flows can be analysed using the experience of self-aerated flows on smooth and rockfill spillways (paragraph 4.4.1). Although the flow conditions are different, the mechanisms of air entrainment are similar to those observed on stepped chutes with skimming flow (fig. 4-1). Once the flow becomes fully developed, the skimming flow behaves in the same way as a smooth or rockfilled chute flow.

For uniform aerated flows, the momentum equation (eq. (4-19)) yields :

$$f_e = \frac{8 * g * \sin\alpha * d_o^2}{q_w^2} * \frac{D_H}{4} \qquad (4-34)$$

where d_o is the water flow depth (eq. (4-14)) in uniform aerated flow.

For the large values of the Reynolds number and the relatively large roughness of spillways, the effects of the Reynolds number can be neglected. If f is the friction factor of non-aerated flow, dimensional analysis (WOOD 1983) suggests that equation (4-25) can be rewritten in the form :

$$\frac{f_e}{f} = f_6\left(C_{mean}; \frac{k_s}{D_H}\right)$$

(4-35a)

For uniform flows, C_{mean} equals C_e and it becomes :

$$\frac{f_e}{f} = f_6\left(C_e; \frac{k_s}{D_H}\right)$$

(4-35b)

The author re-analysed experimental data obtained on smooth chutes (table 4-4). For these data, the drag reduction (eq. (4-35)) is estimated as :

$$\frac{f_e}{f} = 0.5 * \left(1 + \tanh\left(0.628 * \frac{0.514 - C_e}{C_e * (1 - C_e)}\right)\right)$$

(4-36)

where $\tanh(x) = (e^x - e^{-x})/(e^x + e^{-x})$. Experimental data for smooth chutes are presented in figure 4-19 and compared with equation (4-36). Figure 4-19 indicates that the aerated friction factor f_e departs from the non-aerated value f for mean air concentration larger than 20%. In uniform aerated flow, a value $C_e = 0.2$ corresponds to a channel slope of 12 degrees (fig. 4-8, table 4-5).

The study of HARTUNG and SCHEUERLEIN (1970) on rockfilled channels took into account the air entrainment on the flow resistance calculations. The extremely rough bottom induced a highly turbulent flow with air entrainment, and the re-analysis of their results can be presented as :

$$\frac{f_e}{f} = \frac{1}{(1 - 3.2 * \sqrt{f} * \text{Log}_{10}(1 - C_e))^2}$$

(4-37)

Fig. 4-20 - Drag reduction caused by free-surface aeration on rockfill channels
HARTUNG and SCHEUERLEIN (1970) (eq. (4-37)

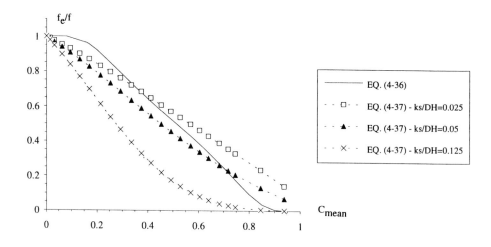

where C_e is estimated from equation (4-18) and f is deduced from equation (4-33). Equation (4-37) is plotted on figure 4-20 for several relative roughnesses. Figure 4-20 shows also a reduction in the ratio f_e/f with an increase in air concentration. Further in fully rough turbulent flows, equation (4-37) suggests that the ratio f_e/f decreases with increasing roughness.

Figures 4-19 and 4-20 indicate a similar trend on both smooth and extremely rough channels; that is, a substantial drag reduction when the air concentration increases above 20%. It is believed that the same process of drag reduction applies also to stepped channels. If the non-aerated friction factor is known, equations (4-36) and (4-37) can be used to provide an estimate of the friction factor of aerated flows on stepped chutes.

4.6 Velocity distribution

Some researchers (FRIZELL 1992, TOZZI 1992) performed velocity measurements for channel slopes of 27 and 53.1 degrees with horizontal steps, in the gradually varied flow region. The analysis of the data indicates that, in both cases, the velocity distribution follows a power law :

$$\frac{V}{V_{max}} = \left(\frac{y}{d}\right)^{1/N}$$
(4-38)

where V_{max} is the velocity near the free-surface. The exponent of the velocity distribution is about : N = 3.5 and 4 for the experiments of FRIZELL and TOZZI respectively.

In uniform non-aerated flows above a smooth chute, CHEN (1990) derived a theoretical relationship between the exponent N of the velocity distribution and the friction factor as

$$N = K * \sqrt{\frac{8}{f}}$$
(4-39)

where K is the Von Karman constant (K = 0.4). For the value of N obtained with FRIZELL's (1992) data, equation (4-39) predicts : f = 0.10. Such a value is of the same order of magnitude as the experimental values deduced from equation (4-27) : that is, f in the range 0.09 up to 0.18 (fig. 4-10).

Discussion

For free-surface aerated flows on smooth chutes, prototype and model data indicate that the velocity distributions follow also a power law (CHANSON 1993a) :

$$\frac{V}{V_{90}} = \left(\frac{y}{Y_{90}}\right)^{1/N}$$
(4-40)

where V_{90} is the characteristic velocity at Y_{90}. Equation (4-40) applies to both uniform and gradually-varied flows, and the exponent N is independent of the mean air concentration C_{mean}.

On stepped chutes, self-aerated skimming flows are expected to behave as free-surface aerated flows on smooth spillways. It is believed that the shape of the velocity distribution is affected by

the large flow resistance. For a smooth concrete chute (i.e. Aviemore spillway), CHANSON
(1993a) obtained : N = 6.0. On a stepped channel, the exponent N is expected to be smaller, and
equation (4-39) might provide an estimate of the exponent N. Typical values are expected to be
close to N = 3.5 to 4 observed on model experiments (FRIZELL 1992, TOZZI 1992).

For known air concentration distributions, the characteristic velocity V_{90} may be deduced from
the continuity equation for the water phase as (CHANSON 1989b) :

$$\frac{V_{90}}{U_w} = \frac{1}{1 - C_{mean}} * \left(\int_0^1 (1 - C) * y'^{1/N} dy' \right)^{-1} \qquad (4\text{-}41)$$

It is interesting to note that BATHURST (1988) and FERRO and BAIAMONTE (1994) measured
velocity distributions in rockfill channel flows. The velocity distributions exhibited a flat S-shape
which can be fitted by equation (4-38) with an exponent N slightly above unity.

4.7 Examples of application

4.7.1 Example No. 1 : Flat slope chute

Considering a stepped chute with a slope of 15 degrees, compute the flow characteristics at the
end of the channel for a water discharge of 25 m^2/s. The dam height is 50 m and the step height
is 1 m.

Calculations

Assuming uniform flow conditions at the end of the channel, equations (4-1) to (4-39), enable to
compute the flow characteristics. For a skimming flow, the boundary layer calculations provide
the flow characteristics and the position of the inception point of air entrainment. The flow
resistance calculations enable to deduce the uniform flow characteristics at the end of the chute.
The main results are given in table 4-8.

The ratio d_c/h equals 4.0. For that value and for $\tan\alpha = h/l = 0.27$, figure 4-3 indicates that the
flow regime is a skimming flow. As the channel slope is flat (i.e. less than 27 degrees), the flow
resistance is estimated using equation (4-30) (sub-regime SK1).

Discussion

Equation (4-1) implies a skimming flow regime. The channel has a flat slope : i.e., $\alpha < 27$ degrees.
At the end of the channel, the non-aerated friction factor is f = 0.31 (eq. (4-30)) and the mean air
concentration is 21% (eq. (4-17)). Taking into account the flow aeration, the aerated flow friction
factor is $f_e = 0.28$ (eq. (4-36)). The application of the momentum equation provides the
knowledge of the flow depth and flow velocity. At the downstream end of the channel, the air-
water flow depth Y_{90} is 2.6 m and the mean flow velocity $(U_w)_o$ is 12.1 m/s.

Table 4-8 -Application for h = 1 m, q_w = 25 m^2/s, α = 15 degrees, H_{dam} = 50 m

Variable (1)	Value (2)	Unit (3)	Eq. (4)	Comments (5)
d_c/h	4.0			Dimensionless critical depth.
$(d_c)_{onset}/h$	0.93		(4-1)	Implies a skimming flow regime.
L_I	62.3	m	(4-11)	Location of the point of inception.
d_I	2.16	m	(4-12)	Flow depth at the point of inception.
C_e	0.21		(4-17)	Average equilibrium air concentration.
d_o	2.06	m	(4-23)	Uniform aerated flow depth.
$(Y_{90})_o$	2.60	m	(4-24)	Characteristic depth where C = 90%.
f	0.31		(4-30)	Non-aerated friction factor (sub-regime SK1).
f_e	0.28		(4-36)	Aerated flow friction factor.
$(U_w)_o$	12.1	m/s	(4-16)	Mean flow velocity.

4.7.2 Example No. 2 : Steep slope chute

Compute the flow characteristics at the end of a stepped channel for a water discharge of 25 m^2/s with a slope of 45 degrees. The dam height is 100 m and the step height is 0.3 m.

After verifying the existence of a skimming flow, the location of the inception point of air entrainment is deduced from the boundary layer calculations (eq. (4-11) and (4-12)). For a steep slope (i.e. α > 27 degrees), figure 4-10 shows a broad scatter of point with no apparent correlation. It is suggested to assume f = 1.0 which is the mean value observed with steep slope data (paragraph 4.5.2). The flow characteristics at the end of the chute are computed assuming uniform flow. The main results are given in table 4-9.

In summary, equation (4-1) implies a skimming flow regime as in the previous example. We observe that the flow is fully-developed before the end of the channel (L_I = 52.4 m). All uniform flow calculations are performed assuming f = 1.0 (paragraph 4.5.2).

Table 4-9 -Application for h = 1 m, q_w = 25 m^2/s, α = 45 degrees, H_{dam} = 100 m

Variable (1)	Value (2)	Unit (3)	Eq. (4)	Comments (5)
d_c/h	13.3			Dimensionless critical depth.
$(d_c)_{onset}/h$	0.59		(4-1)	Implies a skimming flow regime.
L_I	52.4	m	(4-11)	Location of the point of inception.
d_I	1.30	m	(4-12)	Flow depth at the point of inception.
C_e	0.57		(4-17)	Average equilibrium air concentration.
d_o	1.70	m	(4-23)	Uniform aerated flow depth.
$(Y_{90})_o$	3.91	m	(4-24)	Characteristic depth where C = 90%.
f	1			Non-aerated friction factor (sub-regime SK3). Mean value of observed data (fig. 4-10).
f_e	0.43		(4-36)	Aerated flow friction factor.
$(U_w)_o$	14.7	m/s	(4-16)	Mean flow velocity.

CHAPTER 5
SKIMMING FLOW REGIME (2)
ENERGY DISSIPATION

5.1 Presentation

In a skimming flow regime, the water flow exhibits large friction losses over the stepped bottom. Most of the energy is dissipated in maintaining recirculation vortices in the cavities beneath the pseudo-bottom formed by the step edges (fig. 1-3(B), 4-1 and 4-11). If uniform flow conditions are reached before the end of the chute, analytical calculations of the energy dissipation can be developed. In this chapter, such calculations are presented. The results are compared with experimental data. The energy dissipation characteristics of nappe and skimming flows are compared also.

5.2 Head loss calculations

Assuming that the flow is uniform at the downstream end of the channel, the total head loss equals :

$$\frac{\Delta H}{H_{max}} = 1 - \frac{\dfrac{d_o}{d_c} * \cos\alpha + \dfrac{1}{2} * E_C * \left(\dfrac{d_c}{d_o}\right)^2}{\dfrac{3}{2} + \dfrac{H_{dam}}{d_c}} \qquad \text{Un-gated chute (5-1a)}$$

$$\frac{\Delta H}{H_{max}} = 1 - \frac{\dfrac{d_o}{d_c} * \cos\alpha + \dfrac{1}{2} * E_C * \left(\dfrac{d_c}{d_o}\right)^2}{\dfrac{H_{dam} + H_o}{d_c}} \qquad \text{Gated chute (5-1b)}$$

where H_{max} is the maximum head available, H_{dam} is the dam crest head above the downstream toe, H_o is the reservoir free-surface elevation above the chute crest, d_o is the uniform flow depth, d_c is the critical flow depth and E_C is the kinetic energy correction coefficient (Coriolis coefficient). For a velocity power law (eq. (4-38)), the Coriolis coefficient equals :

$$E_C = \frac{(N+1)^3}{N^2 * (N+3)}$$

where $1/N$ is the velocity distribution exponent (eq. (4-38)).

For an un-gated channel, the maximum head available and the weir height are related by : $H_{max} = H_{dam} + 1.5*d_c$. For a gated chute, the relationship becomes : $H_{max} = H_{dam} + H_o$.

Replacing the uniform flow depth by its expression (eq. (4-23)), the head loss is rewritten in terms of the friction factor, the channel slope, the critical depth and the dam height :

$$\frac{\Delta H}{H_{max}} = 1 - \frac{\left(\frac{f_e}{8 * \sin\alpha}\right)^{1/3} * \cos\alpha + \frac{E_C}{2} * \left(\frac{f_e}{8 * \sin\alpha}\right)^{-2/3}}{\frac{3}{2} + \frac{H_{dam}}{d_c}} \qquad \text{Un-gated chute (5-2a)}$$

$$\frac{\Delta H}{H_{max}} = 1 - \frac{\left(\frac{f_e}{8 * \sin\alpha}\right)^{1/3} * \cos\alpha + \frac{E_C}{2} * \left(\frac{f_e}{8 * \sin\alpha}\right)^{-2/3}}{\frac{H_{dam} + H_0}{d_c}} \qquad \text{Gated chute (5-2b)}$$

where f_e is the friction factor of uniform aerated flows (table 4-5, eq. (4-36)).

For a high dam, the residual energy term is small and equation (5-2) is similar to the expression obtained by STEPHENSON (1991) :

$$\frac{\Delta H}{H_{max}} = 1 - \left(\left(\frac{f_e}{8 * \sin\alpha}\right)^{1/3} * \cos\alpha + \frac{1}{2} * E_C * \left(\frac{f_e}{8 * \sin\alpha}\right)^{-2/3}\right) * \frac{d_c}{H_{dam}} \qquad (5-3)$$

Equations (5-2) and (5-3) indicate that the energy dissipation increases with the height of the dam. For high dams however, it becomes more appropriate to consider the residual head H_{res} than the total head loss (see paragraph 5.3).

Discussion

The author has computed the energy dissipation (eq. (5-2a)) for a slope (α = 52 degrees) close to the geometries used by several researchers (table 1-1), using two values of the non-aerated friction factor : f = 0.03 and f = 1.0, that represent the average flow resistance on smooth spillways and on stepped chutes respectively (paragraph 4.5.2). The calculations were performed : 1- neglecting air entrainment (i.e. f_e = f) and 2- taking into account the flow aeration (eq. (4-36)). The results are plotted on figure 5-1 assuming E_C = 1. They are compared with experimental data (table 1-1).

Figure 5-1 indicates a good agreement between the experimental data and equation (5-2) computed with a friction factor f = 1.0 and α = 52 degrees. A comparison between the energy dissipation on smooth and stepped chutes shows clearly a larger energy dissipation occurring along stepped channels for the same discharge and weir height.

Further figure 5-1 shows that the rate of energy dissipation on smooth chutes is affected much more by air entrainment than on stepped channels. As the mean air concentration increases with the slope and the friction factor decreases with the mean air concentration, the effects of air entrainment are more significant on steep slopes. On stepped chutes, air entrainment seems to have little effects on the energy dissipation. But it is more appropriate to consider the residual energy (CHANSON 1993a).

Fig. 5-1 - Energy dissipation in skimming flow regime : comparison between equation (5-2) and experimental data (table 1-1)

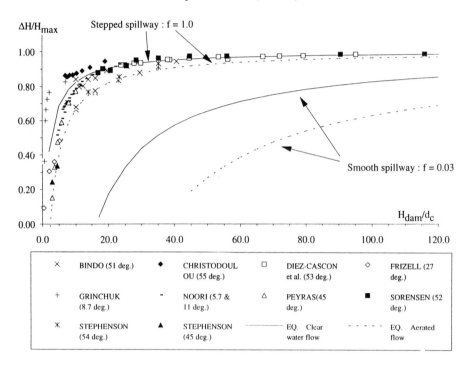

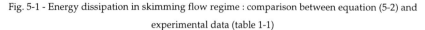

✕	BINDO (51 deg.)	◆	CHRISTODOUL OU (55 deg.)	☐	DIEZ-CASCON et al. (53 deg.)	◇	FRIZELL (27 deg.)
+	GRINCHUK (8.7 deg.)	-	NOORI (5.7 & 11 deg.)	△	PEYRAS(45 deg.)	■	SORENSEN (52 deg.)
✕	STEPHENSON (54 deg.)	▲	STEPHENSON (45 deg.)	——	EQ. Clear water flow	- - - - -	EQ. Aerated flow

5.3 Residual energy

The residual energy is the energy of the flow at the end of the channel. Usually, the residual energy must be dissipated in a dissipation basin at the downstream end of the chute (fig. 5-2). At the bottom of the chute, the residual head H_{res} equals :

$$\frac{H_{res}}{d_c} = \left(\frac{f_e}{8 * \sin\alpha}\right)^{1/3} * \cos\alpha + \frac{1}{2} * E_C * \left(\frac{f_e}{8 * \sin\alpha}\right)^{-2/3} \qquad (5-4)$$

It must be noted that figures (4-19) and (4-20) have shown that the aeration of the flow reduces the flow resistance and increases the kinetic energy of the flow. As a result, the residual energy increases with the amount of entrained air. The relative increase of residual energy caused by the aeration of the flow is :

$$\frac{\Delta(H_{res})}{H_{res}} = \left(\frac{f_e}{f}\right)^{1/3} * \left(\frac{1 + \dfrac{4 * E_C}{f} * \tan\alpha * \left(\dfrac{f}{f_e}\right)}{1 + \dfrac{4 * E_C}{f} * \tan\alpha}\right) - 1 \qquad (5-5)$$

Fig. 5-2 - Stepped spillway model with downstream stilling basin (Courtesy of Dr YASUDA, Nihon University) - $\alpha = 45$ degrees, $q_w = 0.009$ m^2/s, W = 0.4 m

(A) General view of the model and the hydraulic jump at the toe of the stepped channel

(B) Details of the spillway toe and downstream hydraulic jump

Fig. 5-3 - Effects of air entrainment and channel slope on the residual energy (eq. (5-5)) :

$\Delta(H_{res})/H_{res}$ as a function of the mean air concentration C_e and the channel slope α

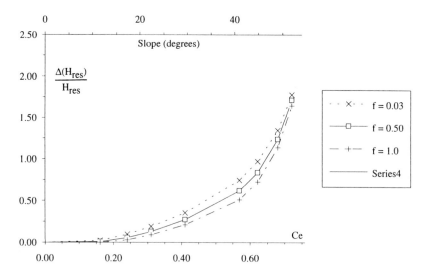

where f_e/f is estimated from equation (4-36) and f is the friction factor of non-aerated flow. On figure 5-3, equation (5-5) is plotted as a function of the mean air concentration and of the channel slope : 1- for a smooth chute (i.e. f = 0.03) and 2- for a stepped channel (i.e. f = 1.0).

Figure 5-3 shows that the residual energy is affected by the flow aeration for mean air concentrations larger than 40%. Figures 4-8 and 5-3, and table 4-5, indicate that a mean air concentration of 40% is obtained for a slope : α = 30 degrees. These results suggest that the effects of air entrainment on the residual energy become important for slopes larger than 30 degrees on both smooth and stepped chutes.

Although figure 5-1 suggests that the total energy dissipation is only slightly overestimated by neglecting the effects of air entrainment, figure 5-3 shows that the residual energy is strongly underestimated if the effect of air entrainment is neglected. It is most important that design engineers take into account flow aeration to estimate the residual energy and to dimension stilling basins downstream of stepped chutes.

5.4 Discussion

5.4.1 Presentation

The total energy dissipation above a long stepped chute is a function of the friction factor f_e, the channel slope α, the discharge and the dam height. The aerated friction factor is a function of the discharge, the channel slope and the slope. In dimensionless terms, the energy dissipation becomes a function of :

$$\frac{\Delta H}{H_{max}} = f_1\left(\alpha ; f ; \frac{H_{dam}}{d_c}\right)$$ (5-6)

It must be emphasised that the calculations depend critically upon the estimation of the friction factor. Figures 4-9 and 4-10 have shown a large scatter of friction factor values observed on model. Therefore equation (5-2) must be used with caution.

Further table 5-1 presents a comparison between the energy dissipation in skimming flows (eq. (5-2a)) for various slopes and dam heights, and taking into account or not the effects of air entrainment (eq. (4-36)). The results indicate clearly that the flow aeration on steep chutes reduces the rate of energy dissipation. For a 60-degree slope, the energy dissipation can be reduced by as much as 20%.

5.4.2 Comparison with smooth chute flows

On a stepped channel, the friction factor is larger than the values observed on a smooth spillway. Such a large increase of friction factor observed on stepped chute will reduce the flow velocity (eq. (4-22)), increase the flow depth (eq. (4-23)) and enhance the energy dissipation (eq. (5-2)). Figure 5-1 shows clearly a larger energy dissipation occurring on stepped chutes than on smooth chutes.

Table 5-1 - Energy dissipation on un-gated stepped spillways

$\dfrac{H_{dam}}{d_c}$	$\dfrac{\Delta H}{H_{max}}$								
	Nappe flow (1)			Skimming flow (2)					
	$\dfrac{d_c}{h}$=0.05	$\dfrac{d_c}{h}$=0.1	$\dfrac{d_c}{h}$=0.5	Un- aerated flow ($f_e = f$)			Aerated flow		
				30 deg.	45 deg.	60 deg.	30 deg.	45 deg.	60 deg.
	Eq. (3-16a)	Eq. (3-16a)	Eq. (3-16a)	Eq. (5-2a)	Eq. (5-2a)	Eq. (5-2a)	Eq. (5-2a)	Eq. (5-2a)	Eq. (5-2a)
(1)	(2)	(3)	(4)	(5)	(6)	(7)	(8)	(9)	(10)
10	0.20	0.45	0.74	0.84	0.83	0.82	0.81	0.70	0.61
20	0.57	0.70	0.86	0.92	0.91	0.90	0.90	0.84	0.79
30	0.71	0.80	0.91	0.94	0.94	0.93	0.93	0.89	0.86
40	0.78	0.85	0.93	0.96	0.95	0.95	0.95	0.92	0.89
50	0.82	0.88	0.94	0.965	0.961	0.96	0.96	0.93	0.91
60	0.85	0.90	0.95	0.971	0.968	0.966	0.96	0.94	0.93
70	0.87	0.91	0.96	0.975	0.972	0.971	0.97	0.95	0.94
80	0.89	0.92	0.964	0.978	0.976	0.974	0.973	0.96	0.95
90	0.90	0.93	0.968	0.980	0.978	0.977	0.976	0.963	0.95
100	0.91	0.937	0.971	0.982	0.980	0.980	0.979	0.966	0.96
120	0.92	0.948	0.976	0.985	0.984	0.983	0.982	0.972	0.963
150	0.94	0.958	0.980	0.988	0.987	0.986	0.986	0.977	0.971
200	0.955	0.968	0.985	0.991	0.990	0.990	0.989	0.983	0.978

Notes : (1) The number of steps equals : $N_{step} = H_{dam}/d_c * d_c/h$.

(2) Calculations made assuming : f = 1.0.

5.4.3 Comparison between nappe and skimming flow regimes

Several researchers (ELLIS 1989, PEYRAS et al. 1991, CHAMANI and RAJARATNAM 1994) suggested that there is much higher energy dissipation in nappe flows than in skimming flow situations.

Figure 5-4(A) and table 5-1 compare the rate of energy dissipation in nappe flows (eq. (3-16a)) and skimming flows (eq. (5-2a)). On figure 5-4(A), equation (3-16a) is plotted for 1, 5 and 10 steps, and equation (5-2a) is presented for slopes ranging from 30 to 70 degrees, assuming f = 1.0. Experimental data obtained with skimming flow situations are also shown. For large dams (i.e. $H_{dam}/d_c > 35$), figure 5-4(A) indicates clearly that a skimming flow regime can dissipate higher flow energy than nappe flow regime. Note also that the channel slope has little effect on the rate of energy dissipation for f = 1.0.

Figure 5-4(B) compares experimental data with nappe and skimming flows. It shows that larger energy dissipation is achieved with skimming flow regime. Further, nappe flow data (HORNER 1969) show that the residual energy (i.e. $H_{res} = 1 - \Delta H/H_{max}$) is 50 to 100% larger than for the skimming flow data (fig. 5-4(B)).

Table 5-1 indicates that, if the spillway is long enough (i.e. if uniform flow conditions are obtained), and for identical flow conditions and dam height, the maximum energy dissipation along the spillway is obtained for a skimming flow regime above a 30-degree slope stepped spillway. On steeper chutes, the flow aeration reduces the flow resistance and hence the rate of energy dissipation.

For short channels, the uniform flow conditions are not obtained at the toe of the waterway. Equation (5-2) becomes inaccurate and overestimates the energy dissipation in skimming flows. In a nappe flow regime, energy dissipation takes place at each step. It is believed that nappe flow situations can dissipate higher energy than skimming flow regime on short chutes. It must be noted however that, for a given discharge, a nappe flow regime requires flatter slope and larger steps (eq. (4-1), fig. 4-3) than a skimming flow regime. In some cases, such requirements might increase the cost of the structure or are not possible.

5.4.4 Superposition of large steps

On the model of the Kennedy's Vale dam ($\alpha = 45$ degrees), STEPHENSON (1988) superimposed large steps to smaller steps. For this model, his findings indicated that occasional large steps induced an additional 10% of specific energy dissipation.

Further experimental investigations are required to verify this finding. In any case, occasional large steps might induce shock waves and flow disturbances that are not acceptable. Designers should not consider the introduction of occasional large steps without new extensive model studies.

Fig. 5-4 - Comparison of energy dissipation in nappe flow regime and skimming flow regime

(A) Comparison of energy dissipation (nappe flow : eq. (3-16a) and skimming flow : eq. (5-2a))

and skimming flow data (table 1-1)

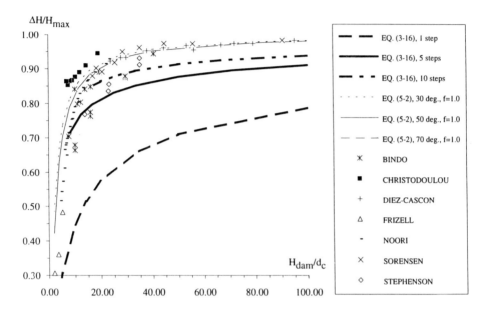

(B) Comparison between nappe flow data (HORNER 1969) and skimming flow data (table 1-1)

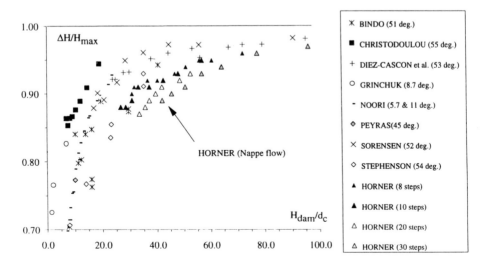

5.5 Examples of application

5.5.1 Example No. 1

For a given discharge, dam height, channel slope and step height, compute the flow conditions at the end of the chute and the residual energy. The dam height is 60 m, the water discharge is 25 m^2/s, the slope is 45 degrees and the step height is 1 m.

The main results are summarised in table 5-2.

Discussion

In summary, the water flows as a skimming stream (eq. (4-1)) above a steep slope (i.e. $\alpha > 27$ deg., sub-regime SK3). The inception point of air entrainment is located 48 m downstream of the spillway crest (eq. (4-11)) The flow depth at that location is 1.5 m (eq. (4-12)).

At the downstream end of the channel (i.e. 85 m downstream of the crest), the non-aerated friction factor can be estimated as 1.0 (fig. 4-10, paragraph 4.5.2). The mean equilibrium air concentration is 57% (fig. 4-8, table 4-5) and the drag reduction induced by air entrainment is nearly 56% (fig. 4-19, table 4-5).

The main uniform flow characteristics are deduced from the momentum equation (paragraph 4.4.2). The aerated flow depth $(Y_{90})_o$ is 3.9 m and the flow velocity $(U_w)_o$ is 14.7 m/s.

The residual head at the downstream end is 12.3 m (eq. (5-4)) and the rate of energy dissipation along the chute is 81% (eq. (5-2)).

The residual hydraulic power is defined as :

$$P_{res} = \rho_w * g * Q_w * H_{res} \qquad\qquad (5\text{-}7)$$

where P_{res} is the hydraulic power and Q_w is the water discharge. At the end of the spillway, the residual power is 3000 kW per unit width.

Table 5-2 -Application for h = 1 m, q_w = 25 m^2/s, α = 45 degrees, H_{dam} = 60 m

Variable (1)	Value (2)	Unit (3)	Eq. (4)	Comments (5)
d_c/h	4.0			Dimensionless critical depth.
$(d_c)_{onset}/h$	0.59		(4-1)	Implies a skimming flow regime.
L_I	48.2	m	(4-11)	Location of the point of inception.
d_I	1.49	m	(4-12)	Flow depth at the point of inception.
C_e	0.57		(4-17)	Average equilibrium air concentration.
d_o	1.70	m	(4-23)	Uniform flow depth.
$(Y_{90})_o$	3.91	m	(4-24)	Characteristic depth where C = 90%.
f	1.0			Non-aerated friction factor (sub-regime SK3).
f_e	0.43		(4-36)	Aerated flow friction factor.
$\Delta H/H_{max}$	0.81		(5-2)	Rate of energy dissipation.
H_{res}	12.3	m	(5-4)	Residual head at the end of the chute.
P_{res}/W	3E+6	W/m	(5-7)	Residual hydraulic power per unit width.

5.5.2 Example No. 2

Considering a flat spillway (α = 12 degrees) and a dam height of 40 m, select the optimum step height for a discharge of 21 m^2/s such as : 1- a skimming flow regime occurs and 2- the residual energy at the downstream end of the spillway is minimum.

The critical depth for the onset of skimming flow is : $(d_c)_{onset}/h$ = 0.6. This value gives a step height of 6 m. For a smaller step height, a typical example of calculations is presented in table 5-3 (i.e. for h = 2 m). As the channel slope is flat (i.e. α < 27 deg., sub-regime SK1), the non-aerated friction factor is estimated with equation (4-30).

For step heights larger than 3.5 m (i.e. d_c/h > 1), the risks of transition between nappe low and skimming flow become important. The flow conditions near the transition nappe/skimming-flow are characterised by transitory fluctuations which might induce improper or dangerous flow behaviours. For that reason, it is decided to select a step height less than 3.5 m.

Complete calculations were performed for step heights ranging from 0.3 up to 3.5 m. The main results are summarised in table 5-4 : they indicate that the residual energy decreases with increasing step heights. To minimise the residual energy and to prevent the apparition of nappe flow, a step height of 3 m would be selected. At the end of the spillway, the residual hydraulic power is 1.18 MW per metre width of channel for h = 3 m.

Table 5-3 -Calculations for h = 2 m, q_w = 21 m^2/s, α = 12 degrees, H_{dam} = 40 m

Variable (1)	Value (2)	Unit (3)	Eq. (4)	Comments (5)
d_c/h	1.8			Dimensionless critical depth.
$(d_c)_{onset}/h$	0.59		(4-1)	Implies a skimming flow regime.
L_I	55.6	m	(4-11)	Location of the point of inception.
d_I	2.27	m	(4-12)	Flow depth at the point of inception.
C_e	0.17		(4-17)	Average equilibrium air concentration.
d_o	2.36	m	(4-23)	Uniform flow depth.
$(Y_{90})_o$	2.83	m	(4-24)	Characteristic depth where C = 90%.
f	0.51		(4-30)	Non-aerated friction factor (sub-regime SK1).
f_e	0.49		(4-36)	Aerated flow friction factor.
$\Delta H/H_{max}$	0.88		(5-2)	Rate of energy dissipation.
H_{res}	6.34	m	(5-4)	Residual head at the end of the chute.
P_{res}/W	1.3E+6	W/m	(5-7)	Residual hydraulic power per unit width.

Table 5-4 -Calculations for q_w = 21 m^2/s, α = 12 degrees, H_{dam} = 40 m

h	H_{res}	P_{res}/W	Comments
m	m	W/m	
(1)	(2)	(3)	(4)
0.3	10.6	2.7E+6	
0.5	8.54	2.2E+6	
1	6.44	1.7E+6	
1.5	5.65	1.4E+6	
2	5.34	1.3E+6	
2.5	5.26	1.22E+6	
3	5.3	1.18E+6	Design step height.
3.5	5.43	1.14E+6	

CHAPTER 6
AIR-WATER GAS TRANSFER ON STEPPED CHUTES

6.1. Introduction

6.1.1 Presentation

One of the most important water quality parameters in rivers and streams is the dissolved oxygen content (DOC). The oxygen concentration is a prime indicator of the quality of the water. Dams and weirs across a stream affect the air-water gas transfer dynamics. Deep and slow pools of water upstream of a dam reduce the gas transfer process and the natural re-aeration as compared to an open river. But some hydraulic structures (e.g. spillway, stilling basin) can enhance the oxygen transfer during water releases.

Environmental applications of stepped chutes and cascades

In rivers, artificial stepped cascades and weirs can be introduced to enhance the dissolved oxygen content of polluted or eutrophic streams (e.g. ERVINE and ELSAWY 1975, AVERY and NOVAK 1978, NAKASONE 1987). In USA, the Tennessee Valley Authority (TVA) designed and built in-stream cascades and drop structures to re-oxygenate rivers (HAUSER et al. 1992, HALL 1993).

With large dams, the downstream nitrogen content is another important parameter. Nitrogen supersaturation may increase the mortality of some fish species. In the Columbia and Snake rivers (USA), 'gas bubble disease' caused by water supersaturated with nitrogen gas was a significant cause of fish mortality for salmonids and steelheads (BOYER 1971, SMITH 1973).

In the treatment of drinking water, cascade aeration can be used to remove gases (e.g. volatile organic compounds VOC) and to eliminate or reduce offensive taste and odour. Inclined corrugated channels and drop structures are effective means for water treatment (BOYDEN et al. 1990, CORSI et al. 1992). The combination of flow aeration and high flow turbulence enhances greatly the mass transfer of VOC's such as chlorine. When power is not required, this process can be successful, since the operation requires little maintenance.

In this chapter, the processes of air-water gas transfer on stepped channels are presented. The flow aeration observed on both nappe flow and skimming flow regimes enhances the gas transfer. And the prediction of dissolved gas contents downstream of a stepped channel is different from smooth channel calculations. In a first part, a review of previous works is presented. Most studies provided empirical correlations to estimate the overall transfer efficiency but the application of such correlations to a new structure is somewhat uncertain. Then a new method to predict the gas transfer caused by self-aeration is developed. The results are discussed and practical applications are developed.

6.1.2 Air-water gas transfer

Fick's law states that the mass transfer rate of a chemical across an interface normal to the x-direction and in a "quiescent" fluid varies directly as the coefficient of molecular diffusion D_{gas} and the negative gradient of gas concentration (STREETER and WYLIE 1981) :

$$\frac{d}{dt} M_{gas} \propto - D_{gas} * \left(\frac{d}{dx} C_{gas} \right) \tag{6-1}$$

where C_{gas} is the concentration of the dissolved chemical in liquid. The dissolved gas concentration gradient, in the fluid layers surrounding a gas bubble, forms a concentration boundary layer. The analysis of this layer for a bubble is very complex because of the bubble shape, the presence of laminar or turbulent flow, a mobile interface in the case of large air bubbles and the interactions between concentration boundary layers from adjacent bubbles.

More generally, the gas transfer of a dissolved chemical (e.g. oxygen, nitrogen) across an air-water interface is written as :

$$\frac{d}{dt} M_{gas} = K_M * A * \left(\frac{P_{gas}}{H_{gas}} - C_{gas} \right) \tag{6-2}$$

where K_M is the mass transfer coefficient, A is the gas-liquid interface area, P_{gas} is the partial pressure of the chemical in air and H_{gas} is the Henry's law constant. If the chemical of interest is volatile (e.g. oxygen, chlorine), the transfer is controlled by the liquid phase (e.g. water) and the coefficient of transfer is almost equal to the liquid film coefficient K_L (i.e. $K_M \sim K_L$). Further Henry's law states that "the weight of any gas that will dissolve in a given volume of a liquid, at constant temperature, is directly proportional to the pressure that the gas exerts above the liquid". In equation form :

$$C_S = \frac{P_{gas}}{H_{gas}} \tag{6-3}$$

where C_S is the concentration of dissolved gas in water at equilibrium (appendix B). Henry's law constant H_{gas} is a function of the salinity, temperature and surfactants. It must be noted that H_{gas} is not exactly constant as the pressure changes. Dividing by the total air-water mixture volume, equation (6-2) becomes :

$$\frac{d}{dt} C_{gas} = K_L * a * (C_S - C_{gas}) \tag{6-4}$$

where a is the specific surface area defined as the air-water interface area per unit volume of air and water.

Discussion

The driving force of the gas transfer is the concentration gradient ($C_{gas} - C_S$). When C_S is greater than C_{gas}, the gas will go into solution (i.e. dissolution). When C_{gas} is greater than C_S (i.e. supersaturation), the gas will desorb.

On stepped chutes, both the aeration of the flow and the strong turbulent mixing enhance the air-water transfer of chemicals. The chemicals can be atmospheric gases (e.g. nitrogen, oxygen, carbon dioxide) or polluted matters (e.g. volatile organic components VOC). The strong turbulent mixing increases the coefficient of transfer K_L in comparison with a quiescent fluid. And the large amount of entrained air bubbles increases the air-water interface area due to the cumulative bubble surface areas. For example, if the bubble diameter is 1 mm and the air content is 10%, the specific interface area is 600 m^2 per cubic metre of air and water.

Application : aeration efficiency of hydraulic structures

If equation (6-4) is integrated along a channel or at a hydraulic structure, the overall gas transfer at cascades and overflow weirs can be measured by the deficit ratio r defined as :

$$r = \frac{C_S - C_{US}}{C_S - C_{DS}} \tag{6-5}$$

where C_{US} is the upstream dissolved gas concentration and C_{DS} is the dissolved gas concentration at the downstream end of the channel. Another measure of aeration is the aeration efficiency E :

$$E = \frac{C_{DS} - C_{US}}{C_S - C_{US}} = 1 - \frac{1}{r} \tag{6-6}$$

Table 6-1 - Temperature dependence of air-water gas transfer

Reference (1)	Temperature dependence (2)	Comments (3)
GAMESON et al. (1958)	$\frac{Ln\,(r_T)}{Ln(r_{15})} = 1 + 0.027*(T - 288.15)$	$273.15 < T < 313.15$ K Oxygen transfer.
ELMORE and WEST (1961)	$\frac{Ln(r_T)}{Ln(r_{20})} = 1.0241^{(T-293.15)}$	$278.15 < T < 303.15$ K Oxygen transfer. Adopted by APHA et al. (1989).
HOLLER (1971)	$\frac{Ln(r_T)}{Ln(r_{T_0})} = \Theta^{(T-T_0)}$	Θ in the range 1.014 to 1.047. Oxygen transfer.
ESSERY et al. (1978)	$(r_T - 1) = (r_{20} - 1)*[1 + 0.0335*(T - 293.15)]$	$283.15 < T < 303.15$ K Oxygen transfer.
DANIIL and GULLIVER (1988)	$\frac{Ln(r_T)}{Ln(r_{20})} = \sqrt{\frac{T}{293}} * \frac{v_{20}}{v} * \sqrt{\frac{\rho_{20}}{\rho}}$	$273.15 < T < 313.15$ K Gas transfer controlled by liquid phase.
GULLIVER et al. (1990)	$\frac{Ln(r_T)}{Ln(r_{T_0})} = \sqrt{\frac{T}{T_0}} * \left(\frac{\mu_{T_0}}{\mu_T}\right)^{3/4} * \left(\frac{\rho_{T_0}}{\rho_T}\right)^{17/20} * \left(\frac{\sigma_{T_0}}{\sigma_T}\right)^{3/5}$	$273.15 < T < 313.15$ K Liquid phase controlled gas transfer.
DANIIL et al. (1991)	$\frac{Ln(r_T)}{Ln(r_{20})} = 1 + 0.02103*(T-273.15) + 8.261E\text{-}5*(T-273.15)^2$	$273.15 < T < 313.15$ K Oxygen transfer.

Notes : T_0 : reference temperature

r_{15} : deficit ratio at 15 Celsius

$r_{20}, v_{20}, \rho_{20}$: deficit ratio, kinematic viscosity and density at 20 Celsius

Effects of temperature

The temperature effect on the air-water gas transfer was examined by several researchers (table 6-1). APHA et al. (1989) suggests to use an exponential relation to describe the temperature dependence on oxygen transfer :

$$\frac{Ln(r)}{Ln(r_{T_0})} = 1.0241^{(T - T_0)} \tag{6-7}$$

where r is the deficit ratio at temperature T, T_0 is a reference temperature and the constant 1.0241 was obtained by ELMORE and WEST (1961).

The formulation of GULLIVER (1990) is more general and applies to gas transfer controlled by the liquid phase (e.g. oxygen, chlorine).

6.1.3 Bibliographic review

6.1.3.1 Gas transfer at stepped chutes

The first measurements of deficit ratio at stepped structures are probably those reported by GAMESON (1957) (table 6-2). This work was extended by GAMESON et al. (1958) who correlated the deficit ratio (for dissolved oxygen content) as :

$$r = 1 + 0.469 * A_G * (1 + 0.046*TC) * \Delta H \tag{6-8}$$

where TC is the temperature of water in Celsius and ΔH is the fall height in metres. The coefficient A_G equals 1.8, 1.6, 1 and 0.65 for clean water, slightly polluted water, moderately polluted water and sewage effluent respectively (Department of Environment 1973). BARRETT et al. (1960) verified equation (6-8) on a short stepped weir. Equation (6-8) was later refined by the Department of Environment (1973) as :

$$r = 1 + (0.38*B) * A_G * (1 + 0.046*TC) * \Delta H * (1 - 0.11*\Delta H) \tag{6-9}$$

where the coefficient B equals 1.30 for the data of GAMESON (1957,1958) and JARVIS (1970) (eq. (6-8)) but varies between 0.7 and 1.3 for model data obtained by TEBBUTT (1972). Note that ΔH is in metres in both equations (6-8) and (6-9). BUTTS and EVANS (1983) correlated equation (6-9) with prototype data and deduced a coefficient B in the range 0.65 to 1.14.

TEBBUTT et al. (1972,1977) and ESSERY et al. (1978) measured aeration efficiency on stepped chute models (table 6-2). ESSERY et al. (1978) proposed the empirical correlation :

$$r_{20} = exp\left(\frac{H_{dam}}{\sqrt{g * h}} * \left(0.427 + 0.31 * Ln\left(\frac{d_c}{h}\right)\right)\right)$$

where H_{dam} is the crest elevation above the channel toe, h is the step height, g is the gravity constant, d_c is the critical flow depth, r_{20} is the deficit ratio at 20 Celsius and all variables are expressed in SI units. The results of TEBBUTT and ESSERY suggest that nappe flow situations with small water discharges provide the best efficiency, and that the step angle δ has little effect on the re-aeration. But the dimensions of their models were small. It is believed that their results cannot be extrapolated to larger structures because they are affected by scale effects.

Photographs by TEBBUTT (1972) indicate clearly that little air entrainment occurred and WOOD (1991) showed that a minimum flow velocity has to be exceeded for the initiation of air entrainment (e.g. paragraph 2.4). The author is convinced that the data of TEBBUTT and ESSERY, obtained with skimming flow regime, underestimate grossly the entrainment of air bubbles and hence the aeration efficiency.

Table 6-2 - Gas transfer experiments on stepped chute and strip roughness

Ref.	Slope deg.	q_w m^2/s	h m	d_c/h	Data	Remarks
(1)	(2)	(3)	(4)	(5)	(6)	(7)
Stepped chutes GAMESON (1957)					Oxygen transfer	H_{dam} = 0.9 to 2.2 m
BARRETT et al. (1960)					Oxygen transfer	H_{dam} = 2.6 m, W = 64 m, T = 25 Celsius. Skimming flow.
TEBBUTT (1972)	35.5	2.8E-4 to 0.00153	0.05	0.002 to 0.124	Oxygen transfer	
	45	0.00056 to 0.013	0.073	0.043 to 0.36	Oxygen transfer	H_{dam} = 1.8 m, W = 0.3 m.
	45	0.00056 to 0.013	0.127 & 0.254	0.0125 to 0.206	Oxygen transfer	H_{dam} = 1.8 m, W = 0.15 m.
TEBBUTT et al. (1977)	11.3, 21.8 & 45	0.077 to 0.965	0.05	1.69 to 9.12	Oxygen transfer	H_{dam} = 2 m, W = 0.15 m. 13 < T < 30 Celsius.
	11.3, 21.8 & 45		0.1	0.85 to 4.56		
	11.3, 21.8 & 45		0.25	0.34 to 1.82		
	11.3, 21.8 & 45		0.5	0.17 to 0.91		
ESSERY et al. (1978)	21.8, 45	0.01 to 0.145	0.025 to 0.5	0.05 to 2.6	Oxygen transfer	W = 0.15 m. Horizontal and inclined upwards steps : δ = 0, 10, 20 deg.. T = 20 Celsius. H_{dam} = 2 m.
	11.3					H_{dam} = 1 m.
BUTTS and EVANS (1983)			0.3048		Oxygen transfer	H_{dam} = 1.22 m. North Aurora and Montgomery dams.
			0.3048		Oxygen transfer	H_{dam} = 2.13 m. McHenry dam.
CIRPKA et al. (1993)		Q_w = 3.3 m^3/s			Transfer of VOC's	Stepped weirs along a natural stream with nappe flow. VOC : SF_6, R113, C_2HCl_3, $CHCL_3$, $CHBr_3$
	6.0		0.2 to 0.5			H_{dam} = 1.72 m. 3 pooled steps (pool height : 0.4 m) and 1 flat step. Trapezoidal channel : W = 18 to 20 m.
	6.6		0.2 to 0.5			H_{dam} = 1.35 m. 2 pooled steps (pool height : 0.4 m) and 1 flat step. Trapezoidal channel : W = 18 to 20 m.
Strip roughness BICUDO and GIORGETTI (1991)	0.03	0.021 to 0.042			Oxygen transfer	W = 0.25 m, k_s = 15 mm, l_s = 15 mm, L_s = 52, 104, 156, 600 mm

Notes :

SF_6 = sulfur hexafluoride; R113 = 1.1.2-trichlorotrifluoroethane; C_2HCl_3 = trichloroethene; $CHCL_3$ = trichloromethane; $CHBr_3$ = tribromomethane

k_s, L_s, l_s strip roughness height, spacing and length (fig. (4-4))

CIRPKA et al. (1993) measured air-water gas transfer of volatile compounds at small in-stream cascades. Their study indicated that the compound properties (e.g. Henry coefficient) affect strongly the gas exchange process. Further they suggested that the air bubble entrainment has little effects for compounds with small Henry coefficients.

6.1.3.2 Gas transfer at free-overfalls and waterfalls

MASTROPIETRO (1968) measured dissolved oxygen contents downstream of small dams (table 6-3) with large tailwater depths (i.e. $d_t > 4$ m). For free-falling weirs of height less than 4.6 m, he obtained the correlation :

$$r = \frac{1}{1 - 0.141 * \Delta H} \tag{6-10}$$

where ΔH is the difference between water levels above and below the weir (i.e. the head loss) in metres. HOLLER (1971) analysed data for three prototype weir structures and gave :

$$r_{20} = 1 + 0.211 * \Delta H \tag{6-11}$$

FOREE (1976) analysed the re-aeration at low level in-streams dams behaving as waterfalls and correlated his data by :

$$r_{25} = \exp(0.525 * \Delta H) \tag{6-12}$$

where r_{25} is the deficit ratio at 25 Celsius.

AVERY and NOVAK (1978) performed a systematic study of oxygenation with overfall models and with large depths of receiving water. For their data, the oxygen transfer is correlated by :

$$r_{15} = 1 + k_{AN} * \left(\frac{\rho_w}{\mu_w} * (D_H)_i * \sqrt{\frac{g * \Delta H}{8}}\right)^{0.53} * \left(\frac{2 * \Delta H}{(D_H)_i}\right)^{0.89} \tag{6-13}$$

where r_{15} is the deficit ratio at 15 Celsius and $(D_H)_i$ is the hydraulic diameter of the free-falling jet at impact. The coefficient k_{AN} is a function of the salinity (table 6-4).

NAKASONE (1987) performed also experiments on weir models and developed an empirical correlation that takes into account the tailwater depth in the receiving pool :

$$Ln(r_{20}) = k_{N1} * \Delta H^{k_{N2}} * q_w^{k_{N3}} * d_t^{0.310} \tag{6-14}$$

where d_t is the tailwater depth in the receiving pool and the coefficients k_{N1}, k_{N2}, k_{N3} are functions of the discharge and drop height. For the large size experiments (i.e. $\Delta H > 1.2$ m and $q_w > 0.065$ m^2/s), equation (6-14) becomes :

Table 6-3 - Free-overfall experiments

Ref.	q_w m^2/s	Fall height m [a]	Data	Remarks
(1)	(2)	(3)	(4)	(5)
Water Pollution (1957)		0.61 & 1.52	Oxygen transfer	Model data. $0 < d_t < 0.305$ m.
MASTROPIETRO (1968)		1.22 to 4.6	Oxygen transfer	Prototype data. $20 < T < 25$ Celsius
HOLLER (1971)		3.7, 8.4, 10.7	Oxygen transfer	Prototype data. Tailwater depths < 4.9 m.
APTED and NOVAK (1973)	0.0062 to 0.021	1.15	Oxygen transfer	Model data. W = 0.1 m. $0.05 < d_t < 0.45$ m.
Department of Environment (1973)		0.15, 0.61, 1.52, 3.05	Oxygen transfer	Model data. $0 < d_t < 0.45$ m.
FOREE (1976)	$Q_w = 0.057$ to 0.4 m^3/s	0.79 to 7.3	Oxygen transfer	Prototype data. $17 < T < 24$ Celsius
AVERY and NOVAK (1978)	0.003 to 0.05	0.25 to 2.1	Oxygen transfer	Rectangular jets. Model data. W = 0.1, 0.22, 0.3 m
NAKASONE (1987)	0.011 to 0.741	0.24 to 1.98	Oxygen transfer	Weirs. Model data. W = 0.2, 0.3 m
	0.027 to 7.8	1.52 to 5.8	Oxygen transfer	Weirs. Prototype data.
CORSI et al. (1992)	$Q_w = 1.08$ to 3.39 L/s	0.5 to 2.3	Transfer of Oxygen & Volatile Organic Compounds	Drop structure. Model data. T = 20 Celsius. VOC : $CDCl_3$, EDB.

Notes :

[a] :   difference between water levels

d_t tailwater depth

$CDCl_3$ = deuterated chloroform; EDB = ethylenedibromide.

Table 6-4 - Coefficients k_{AN} (AVERY and NOVAK 1978)

Water	k_{AN}	Reference
(1)	(2)	(3)
Tap water	0.627E-4	AVERY and NOVAK (1978)
Tap water + 0.3% of $NaNO_3$	0.869E-4	
Tap water + 0.6% of $NaNO_3$	1.243E-4	

$$Ln(r_{20}) = 5.92 * \frac{\Delta H^{0.816}}{q_w^{0.363}} * d_t^{0.310}$$
(6-15)

where ΔH and d_t are in metres and q_w in m^2/s.

NAKASONE's (1987) correlation (eq. (6-14)) was verified with some model and prototype data for discharges between 0.011 and 7.8 m^2/s, fall heights ranging from 0.24 to 5.8 m and tailwater depths in the range 0.25 to 7.5 m.

The results of NAKASONE (1987) (eq. (6-14)) showed an increase in aeration efficiency with an increasing tailwater depth. Other researchers (Water Pollution 1957, JARVIS 1970, APTED and NOVAK 1973, Department of Environment 1973) observed also an increase of the deficit ratio with the depth of receiving water. But they indicated that, for tailwater depths larger than a characteristics value, the deficit ratio reached an upper limit. Indeed the aeration efficiency must become independent of the receiving pool depth, for tailwater depths larger than the maximum penetration depth of the entrained air bubbles.

For oxygen transfer at drop structures, CORSI et al. (1992) observed a linear increase of the coefficient of transfer $[K_M * A]$ (eq. (6-2)) with increasing drop heights or discharges when the other parameter was held constant. Their experiments showed also that the VOC emission of a 2-m high drop structure can be significant.

6.1.3.3 Gas transfer at hydraulic jumps

A hydraulic jump is characterised by the development of large-scale turbulence, energy dissipation and air entrainment. Gas transfer at hydraulic jumps results from the large number of entrained air bubbles and the turbulent mixing in the jump. Several researchers have proposed correlations to predict the oxygen transfer (table 6-5). JOHNSON (1984) developed an empirical fit of prototype data that works well with relatively deep plunge pools. But RINDELS and GULLIVER (1986) showed that JOHNSON's method does not apply to the re-aeration of waters with low DO contents and with shallow plunge pools.

AVERY and NOVAK (1978) and WILHELMS et al. (1981) performed independently model experiments and they obtained a similar correlation (table 6-5) :

$$r_{15} = 1 + \alpha_1 * \left(\frac{q_w}{\sqrt{g * d_1{}^3}} \right)^{\alpha_2} * \left(\frac{q_w}{v_w} \right)^{\alpha_3} \qquad (6\text{-}16)$$

where d_1 is the upstream flow depth, q_w is the discharge per unit with and v_w is the kinematic viscosity of the water.

6.2 Prediction of air-water gas transfer

6.2.1 Introduction

In the past, most researchers (paragraph 6.1.3) attempted to correlate the overall gas transfer at hydraulic structures. Their works did not take into account the level of turbulent mixing nor the air bubble entrainment processes.

More recently, some researchers (GULLIVER et al. 1990, JUN and JAIN 1993) attempted to correlate the "overall transfer coefficient" $[K_L{}^*a]$ to the flow conditions and they assumed $[K_L{}^*a]$ to be a constant across the hydraulic structure. With their reasoning, equation (6-4) might be integrated over the residence time t of the entrained bubbles to give the overall deficit ratio :

Table 6-5 - Oxygen transfer correlations at hydraulic jumps

Reference (1)	Formula (2)	Remarks (4)
HOLLER (1971)	$r_{20} - 1 = 0.0463 * \Delta V^2$	Model experiments : $0.61 < \Delta V < 2.44$ m/s $277.15 < T < 299.15$ K
APTED and NOVAK (1973)	$r_{15} = 10^{(0.24*\Delta H)}$	Model experiments : $2 < Fr_1 < 8$ $q_w = 0.04$ m^2/s
AVERY and NOVAK (1975)	$r_{15} - 1 = 0.023 * \left(\dfrac{q_w}{0.0345}\right)^{3/4} * \left(\dfrac{\Delta H}{d_1}\right)^{4/5}$	Model experiments : $2 < Fr_1 < 9$ $1.45E+4 < Re < 7.1E+4$ $0.013 < d_1 < 0.03$ m $287.15 < T < 291.15$ K $W = 0.10$ m
AVERY and NOVAK (1978)	$r_{15} - 1 = k' * Fr_1^{2.1} * Re^{0.75}$ k' is function of the salinity (table 6-6).	Model experiments : $1.45E+4 < Re < 7.1E+4$ $v_w = 1.143E-6$ m^2/s $W \doteq 0.10$ m
WILHELMS et al. (1981)	$r_{15} - 1 = 4.924E-8 * Fr_1^{2.106} * Re^{1.034}$	Model experiments (W = 0.381 m) based upon Krypton-85 transfer : $1.89 < Fr_1 < 9.5$ $2.4E+4 < Re < 4.3E+4$
JOHNSON (1984)	Empirical fit based on field data from 24 hydraulic structures.	Prototype data. Working well for deep plunge pools.

Notes : ΔV :   difference between the upstream and downstream velocities (m/s)

d_1 :   upstream flow depth (m).

Fr_1 :   upstream Froude number defined as : $Fr_1 = q_w/\sqrt{g^*d_1^3}$

q_w :   discharge per unit width (m^2/s)

Re   Reynolds number defined as : $Re = q_w/v_w$.

r_{15}, r_{20} :   oxygen deficit ratio at 15 Celsius and 20 Celsius.

W :   channel width (m).

ΔH :   head loss in the hydraulic jump (m of water).

Table 6-6 - Coefficients k' (AVERY and NOVAK 1978)

Water (1)	k' (2)	Reference (3)
Tap water	1.0043E-6	AVERY and NOVAK (1978)
Tap water + 0.3% of NaNO$_2$	1.2445E-6	
Tap water + 0.6% of NaNO$_2$	1.5502E-6	

$$r = \exp(K_L * a * t) \tag{6-17}$$

In most situations however, the interface area a, the coefficient of transfer K_L and the product $[K_L*a]$ vary across a hydraulic structure and cannot be assumed constant. Hence equation (6-17) cannot describe accurately the air-water gas transfer taking place at a hydraulic structure (e.g. on a stepped chute).

6.2.2 A novel approach

The gas transfer analysis must be approached in a different way. If the coefficient of transfer K_L and the interface area a are known at any position (x,y,z), equation (6-4) can be integrated locally. At each location (x,y,z) along a streamline, equation (6-4) can be averaged over a small control volume and it yields :

$$\frac{d}{ds} C_{gas}(x,y,z) = \frac{K_L(x,y,z) * a(x,y,z)}{V(x,y,z)} * (C_s(x,y,z) - C_{gas}(x,y,z)) \tag{6-18}$$

where V is the local flow velocity and s is the direction along a streamline. The total air-water gas transfer at a hydraulic structure is then deduced as :

$$C_{DS} - C_{US} = \int_x \int_y \int_z \left(\frac{K_L(x,y,z) * a(x,y,z)}{V(x,y,z)}\right) * [C_s(x,y,z) - C_{gas}(x,y,z)] * dx * dy * dz \tag{6-19}$$

where C_{DS} is the mean downstream dissolved gas content and C_{US} is the mean upstream dissolved gas content.

Saturation concentration C_s

Along a chute, the pressure variations are relatively small, and the liquid density and viscosity can be assumed functions of the temperature only. In most practical applications, the temperature and salinity are constant at a hydraulic structure, and the saturation concentration C_s becomes a constant in equations (6-4), (6-18) and (6-19). If the pressure variations are important, complete calculations of C_s must be used (see appendix B).

Coefficient of transfer K_L

A recent review of the transfer coefficient calculations in turbulent gas-liquid flows showed that K_L is almost constant regardless of the bubble size and the flow situations (KAWASE and MOO-YOUNG 1992). Using HIGBIE's (1935) penetration theory, the transfer coefficient of gas bubbles affected by surface active impurities can be estimated as :

$$K_L = 0.28 * D_{gas}^{2/3} * \left(\frac{\mu_w}{\rho_w}\right)^{-1/3} * \sqrt[3]{g} \qquad (d_{ab} < 0.25 \text{ mm}) \tag{6-20a}$$

$$K_L = 0.47 * \sqrt{D_{gas}} * \left(\frac{\mu_w}{\rho_w}\right)^{-1/6} * \sqrt[3]{g} \qquad\qquad (d_{ab} > 0.25 \text{ mm}) \quad (6\text{-}20b)$$

where D_{gas} is the molecular diffusivity, μ_w and ρ_w are the dynamic viscosity and density of the liquid, d_{ab} is the air bubble diameter and g is the gravity constant. Equation (6-20) was successfully compared with more than a dozen of experimental studies (KAWASE and MOO-YOUNG 1992). Equation (6-20) implies that the coefficient of transfer is a function of the temperature of liquid only (see appendix B). If the temperature is a constant along a chute, K_L becomes a constant in equation (6-18).

For constant saturation concentration and coefficient of mass transfer, the integration of equation (6-18) is simplified. The prediction of the air-water gas transfer along a chute is possible if the velocity V and air-water interface area a are known at every position (x,y,z). Such a method was applied previously to smooth chutes by the author (CHANSON 1993d).

6.2.3 Air-water interface area a

In a turbulent shear flow, the maximum air bubble size is determined by the balance between the capillary force and the inertial force caused by the velocity change over distances of the order of the bubble diameter. The splitting of air bubbles in water occurs for (HINZE 1955) :

$$\frac{\rho_w * v'^2 * d_{ab}}{2 * \sigma} > (We)_c \qquad\qquad\qquad (6\text{-}21)$$

where σ is the surface tension between air and water, d_{ab} is the bubble diameter, v'^2 is the spatial average value of the square of the velocity differences over a distance equal to d_{ab} and $(We)_c$ is critical Weber number for bubble splitting (table 6-7).

In turbulent air-water flows, observations of bubble diameters indicate that the bubble sizes are larger than the Kolmogorov microscale and smaller than the turbulent macroscale. The observations suggest also that the length scale of the eddies responsible for breaking up the bubbles is close to the bubble size (SEVIK and PARK 1973, KUMAR et al. 1989). These eddies lie within the inertial range and are isotropic. It is believed that large scale eddies will carry air bubbles while eddies with length scales smaller than the bubble size do not have enough energy to break up the bubbles. Bubble breakup results from the interactions between bubbles and turbulent vortices of similar length scale.

For vertical bubbly turbulent jets, SUN and FAETH's (1986) computations suggested that the characteristic eddy size ranged from 0.5 to 5 times the bubble diameter. In bubble column flows, AVDEEV et al. (1991) indicated that the momentum mixing length is nearly equal to the bubble diameter at the location of the maximum turbulent shear stresses. Assuming that the maximum bubble diameter is in the order of magnitude of the Prandtl mixing length (CHANSON 1992a), the turbulent fluctuation equals : $v'^2 = (d_m * \omega)^2$ where d_m is the maximum bubble size and ω is the vorticity (LEWIS and DAVIDSON 1982). If the longitudinal acceleration term dV/dx is small, the vorticity equals : $\omega = dV/dy$ where y is the direction perpendicular to the flow direction. It yields :

$$v'^2 \sim \left(\frac{dV}{dy} * d_m\right)^2 \qquad (6\text{-}22)$$

With this assumption, equation (6-21) can be transformed and the maximum bubble size in a turbulent shear flow can be estimated as :

$$d_m \sim \sqrt[3]{\frac{2 * \sigma * (We)_c}{\rho_w * \left(\frac{dV}{dy}\right)^2}} \qquad (6\text{-}23)$$

Experiments showed that the critical Weber number is a constant near unity (table 6-7).

Equation (6-23) satisfies the common sense that the maximum bubble size increases as the velocity gradient and the turbulent shear stress decrease. Further equation (6-23) was compared successfully with experimental observations in free-surface aerated flows (CHANSON 1993d) and in plunging jet flows (CHANSON and CUMMINGS 1992).

Table 6-7 - Critical Weber numbers for the splitting of air bubbles in water flows

Reference (1)	$(We)_c$ (2)	Fluid (3)	Flow situation (4)	Comments (5)
HINZE (1955)	0.585		Two co-axial cylinders, the inner one rotating	Dimensional analysis. Re-analysis of CLAY's (1940) data.
SEVIK and PARK (1973)	1.26	Air bubbles in water	Circular water jet discharging vertically	Experimental data. V in the range 2.1 to 4.9 m/s.
KILLEN (1982)	1.017	Air bubbles in water	Turbulent boundary layer	Experimental data. V in the range 3.66 to 18.3 m/s.
LEWIS and DAVIDSON (1982)	2.35	Air and Helium bubbles in water and Fluorisol	Circular jet discharging vertically	Experimental data. V in the range 0.9 to 2.2 m/s.
PANDIT and DAVIDSON (1986)	1.1	Air bubbles in water	Circular jet discharging vertically	Experimental data. V in the range 0.49 to 1.8 m/s.
EVANS et al. (1992)	0.60	Air bubbles in water	Confined plunging water jet	Experimental data. V in the range 7.8 to 15 m/s. Note : measurements of bubble size outside jet mixing zone.
MIKSIS et al. (1981)	1.615	Bubble in inviscid incompressible fluid	Uniform flow around a bubble	Numerical calculations. Steady potential flow around a bubble of constant internal pressure.
RYSKIN and LEAL (1984)	0.125 to 1.4	Incompressible gas bubble in a liquid of constant density and viscosity	Steady uniaxial extensional flow	Numerical calculations. $(We)_c$ is a function of Re : $(We)_c = 0.125$ for Re = 1, $(We)_c = 1.4$ for Re infinite.
LEWIS and DAVIDSON (1992)	2.35	Cylindrical bubble surrounded by inviscid liquid	Axi-symmetric shear flow	Theoretical value.

High speed photographs (HALBRONN et al. 1953, STRAUB and LAMB 1953) showed that the shape of air bubbles in self-aerated flows is approximately spherical. Plunging jet experiments performed by the author show also that the shape of bubbles is approximately spherical for velocities ranging from 0.5 to 9 m/s. For quasi-spherical bubbles, the specific interface area equals :

$$a = 6 * \frac{C}{d_{ab}} \qquad\qquad (6\text{-}24)$$

where C is the air concentration (i.e. the concentration of undissolved air). If the air bubble size is estimated using equation (6-23), a conservative estimate of the air-water interface area can be deduced :

$$a = 6 * \frac{C}{d_m} \qquad\qquad (6\text{-}25)$$

If the air concentration C and velocity V distributions are known, equations (6-23) and (6-25) enable the calculation of the air-water surface area at any position along a chute. Then equation (6-18) can be integrated along a stepped chute to provide the rate of air-water gas transfer.

6.3 Nappe flow regime

6.3.1 Introduction

In a nappe flow regime, the air-water gas transfer at each step results from the gas transfer at the plunge point and at the downstream hydraulic jump (fig. 6-1). The aeration efficiency at one individual step equals :

$$E_1 = E^{jet} + E^{HJ} * (1 - E^{jet}) \qquad\qquad (6\text{-}26)$$

where E^{jet} is the aeration efficiency of the plunging jet flow and E^{HJ} is the aeration efficiency of the hydraulic jump. For deep pooled steps, most of the gas transfer results from the plunging jet process :

$$E_1 = E^{jet} \qquad\qquad \text{Deep pooled steps} \quad (6\text{-}27)$$

For a series of N_{step} steps, the aeration efficiency of the complete cascade can be expressed in terms of the aeration efficiency of each step :

$$E = 1 - \prod_{i=1}^{N_{step}}(1 - E_i) \qquad\qquad (6\text{-}28)$$

where E_i is the aeration efficiency of the i-th step, estimated using equation (6-26). For a stepped chute with N_{step} steps of equal characteristics, the aeration efficiency of the chute becomes :

$$E = 1 - (1 - E_1)^{N_{step}} \qquad\qquad (6\text{-}29)$$

Fig. 6-1 - Flow aeration in nappe flow regime (sub-regime NA1)

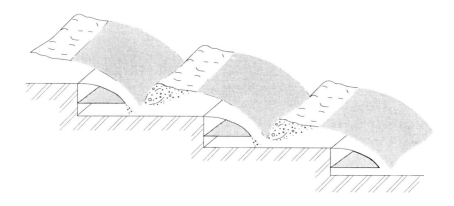

6.3.2 Gas transfer by plunging jet

Air bubble size

The author has developed a simple method to estimate the maximum size of air bubble entrained at the impact of an inclined jet and a receiving pool. The results of the simple model suggests two preferential maximum bubble sizes. At the inner (lower) shear layer of the jet (fig. 6-2), the buoyancy induces bubble trajectories subjected to larger shear stresses than at the outer (upper) shear layer. As a result, smaller bubble size are likely to be observed at the inner edge of the jet.

For an inclined jet at 20 Celsius and atmospheric pressure, the maximum bubble size (eq. (6-23)) can be correlated as :

$$d_m = (We)_c * \left((0.05951 - 0.04069 * (\cos\theta)^{0.6896}) * V_i^{(-2/3 + 0.5075*(\cos\theta)^{0.8732})}\right)^3$$

inner shear layer (6-30)

$$d_m = (We)_c * \left((0.05951 + 0.45462 * (\cos\theta)^{2.0503}) * V_i^{(-2/3 - 1.4931*(\cos\theta)^{1.3853})}\right)^3$$

outer shear layer (6-31)

where θ is the angle of the inclined jet and the receiving pool of water and V_i is the nappe velocity at the intersection with the receiving pool. θ and V_i can be estimated respectively by equations (3-12) and (3-11). Details of the calculations are presented in appendix C.

Maximum bubble penetration depth

A theoretical value of the maximum bubble penetration depth can be deduced from the continuity and momentum equations for diffusing jets. Assuming that the bubbles are entrained to a depth (i.e. maximum penetration depth) where the vertical component of the mean jet velocity equals the bubble rise velocity, CHANSON and CUMMINGS (1992) extended the method of ERVINE and FALVEY (1987) to inclined plane and circular jets. For plane jets, the penetration depth D_p is correlated by:

$$\frac{D_p}{d_i} = 0.0240 * \left(\frac{V_i}{u_r}\right)^2 * \frac{(\sin\theta)^3}{(\tan\theta_3)^2} * \left(1 + \sqrt{1 - 20.81 * \left(\frac{u_r}{V_i}\right)^2 * \frac{\tan\theta_3}{(\sin\theta)^2}}\right)^2 \qquad (6\text{-}32)$$

where D_p is measured normal to the pool free-surface, d_i is the jet thickness, u_r is the bubble rise velocity and θ_3 is the outer spread angle in the fully developed flow region (fig. 6-2). For circular jets, ERVINE and FALVEY (1987) estimated θ_3 around 14 degrees on both models and prototypes.

Fig. 6-2 - Sketch of the flow regions of an inclined plunging jet

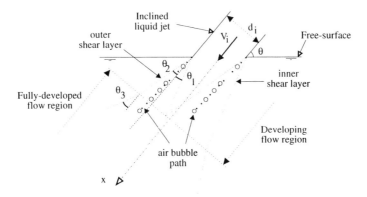

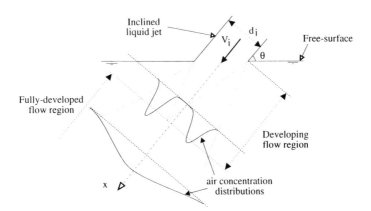

Fig. 6-3 - Air concentration distributions in the downward flow motion region - VAN DE DONK
(1981) : vertical circular jet, d = 6 mm, V = 4 m/s

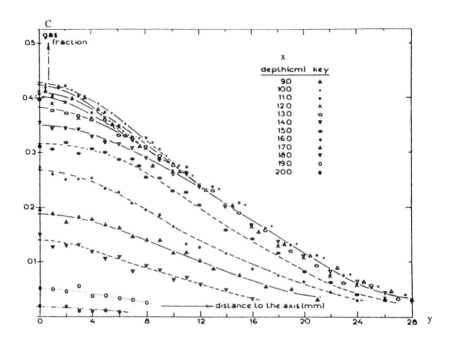

Air bubble diffusion

In the fully-developed flow region of a plunging jet, the re-analysis of experimental data
(McKEOGH and ERVINE 1981, VAN DE DONK 1981, KUSABIRAKI et al. 1990, BONETTO and
LAHEY 1993, author) suggests that the distributions of air bubbles entrained by a plunging jet
follow a Gaussian curve :

$$C = C_{max} * \exp\left(-\left(1.52 * \frac{y - Y_{C_{max}}}{Y_{0.1} - Y_{C_{max}}}\right)^2\right) \tag{6-33}$$

where C_{max} is the maximum air concentration at a cross-section x, x is the direction along the
streamlines and y is the direction normal to the streamlines, $Y_{C_{max}}$ and $Y_{0.1}$ are the locations
were $C = C_{max}$ and $C = 0.1*C_{max}$ respectively. Figure 6-3 shows typical results (VAN DE
DONK 1981).

For vertical circular jets, the re-analysis of the data of VAN DE DONK (1981) and of McKEOGH
and ERVINE (1981) indicate that the decay of the maximum air content can be approximated by
a linear function :

$$C_{max} = (C_{max})_o * \left(1 - \frac{x * \sin\theta}{D_p}\right) \tag{6-34}$$

To a first approximation, the initial maximum air content $(C_{max})_0$ can be deduced from the quantity of air entrained :

$$(C_{max})_0 = \frac{Q_{air}^{jet}/Q_w}{1 + Q_{air}^{jet}/Q_w} \tag{6-35}$$

The re-analysis of a limited set of data (McKEOGH and ERVINE 1981) suggests that the 1/10-th value width of the entrained bubble cloud can be correlated, for a vertical circular jet, by :

$$Y_{0.1} - Y_{C_{max}} = K'' * \left(1 - \frac{x * \sin\theta}{D_p}\right)^{0.33} \tag{6-36}$$

where K'' is a constant.

Equations (6-33) to (6-36) enable to predict the distributions of air concentration in the fully-developed flow region of a plunging liquid jet.

Gas transfer of free-falling jet

The dispersion of entrained air bubbles forms two distinctly different regions : 1- a characteristic bubbly air-water region with a downward flow motion induced by the plunging liquid jet, and 2- a region of rising bubbles which surrounds the former one.

In the bubbly flow region (region 1), equations (6-33) to (6-36) provide an estimate of the air bubble diffusion. The air-water interface area can be deduced from the local air concentration and the maximum bubble size (eq. (6-30) and (6-31)). Recent reviews by BIN (1993) and by the author indicate that there is no information on the velocity profiles and on the momentum diffusion for two dimensional plunging jets (vertical and inclined jet). The lack of knowledge of the velocity distribution prevents the integration of equation (6-18). Presently, active research are under way in Australia and USA (e.g. BONETTO and LAHEY 1993, CUMMINGS 1994). It is believed that these research projects will provide useful information in a near future.

In the swarm of rising bubbles (region 2), the bubble motion is driven by buoyancy. To a first approximation, the air bubble motion is vertical and the bubble velocity equals nearly the bubble rise velocity in still water. Calculations of the bubble rise velocity are detailed in appendix C.

6.3.3 Gas transfer at the hydraulic jump

Introduction

At the toe of a hydraulic jumps, a large amount of air is entrained (eq. (3-26) and (3-27)) and air bubbles are diffused downstream into the roller. Several researchers recorded the distributions of air bubble concentrations at hydraulic jumps with partially developed inflows conditions, fully-developed hydraulic jumps and pre-entrained jumps[1] (table 6-8). Figure 6-4 shows a typical

---

[1]A 'pre-entrained jump' is a fully-developed hydraulic jump with upstream free-surface aeration. The presence of air bubbles within the flow and next to the free-surface modifies the jump characteristics.

example of air concentration distributions along a hydraulic jumps with partially developed inflow conditions.

Structure of the bubbly flow

THANDAVESWARA (1974) presented the most comprehensive analysis of the bubbly flow region of a hydraulic jump. His work was conducted using high-speed photography and conductivity probe measurements. The followings summarise THANDAVESWARA's findings and is consistent with later investigations (table 6-8).

The air-water flow of a hydraulic jump includes three regions : 1- a *turbulent shear layer* with smaller air bubble sizes, 2- a *"boiling" flow region* characterised by the development of large-scale eddies and bubble coalescence, and 3- a *foam layer* at the free-surface with large air polyhedra structures (fig. 6-5).

Air entrainment occurs in the form of air bubbles and air pockets entrapped at the impingement of the upstream jet flow with the roller. The air packets are broken up in very thin air bubbles as they are entrained. When the bubbles are diffused into regions of lower shear stresses, the coalescence of bubbles yields to larger bubble sizes and these bubbles are driven by buoyancy to the boiling region. Near the free-surface, the liquid is reduced to thin films separating the air bubbles. Their shape becomes pentagonal to decahedron as pictured by THANDAVESWARA (1974).

Fig. 6-4 - Air concentration distributions in a hydraulic jump with partially developed inflow : $q_w = 0.0504 \ m^2/s$, $Fr_1 = 8.11$ (CHANSON and QIAO 1994) - x : distance from the jump toe

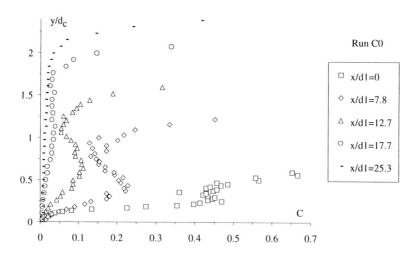

Fig. 6-5 - Air-water flow regions in a hydraulic jump

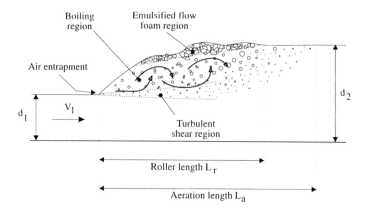

Fig. 6-6 - Distribution of bubble sizes in a fully-developed hydraulic jump - LEUTHEUSSER et al.
(1973) : Fr_1 = 3.26, q_w = 0.03084 m2/s, d_1 = 0.021 m, d_2 = 0.086 m

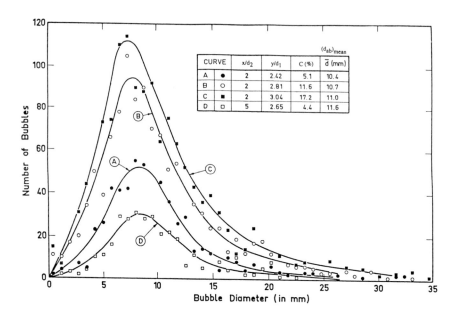

CURVE		x/d_2	y/d_1	C (%)	$(d_{ab})_{mean}$ $\overline{d}$ (mm)
A	●	2	2.42	5.1	10.4
B	○	2	2.81	11.6	10.7
C	■	2	3.04	17.2	11.0
D	□	5	2.65	4.4	11.6

Table 6-8 - Experimental observations of air bubble distributions in hydraulic jumps

Ref.	q_w	Fr_1	Inflow conditions	Comments
	m^2/s		m	
(1)	(3)	(4)	(5)	(6)
RAJARATNAM (1962)	0.0307 to 0.1106	2.42 to 8.72	P/D	W = 0.3048 m.
RESCH and LEUTHEUSSER (1972)	0.0339 and 0.0718	2.98 and 8.04 [a]	P/D	W = 0.39 m.
	0.03084 and 0.0786	3.26 and 7.32 [a]	F/D	W = 0.39 m.
THANDAVESWARA (1974)	0.0302 to 0.06086	7.16 to 13.31	P/D	W = 0.6096 m.
	0.06194 to 0.15577	5.4 to 5.89	PHJ	W = 0.4572 m.
BABB and AUS (1981)	0.12304	6.0	P/D	W = 0.46 m.
CHANSON and QIAO (1994)	0.0312 to 0.0504	5.02 to 8.11	P/D	W = 0.25 m.

Notes :

Inflow conditions : P/D = partially developed upstream flow conditions; F/D = fully-developed upstream flow conditions; PHJ = pre-entrained hydraulic jump.

[a] : RESCH and LEUTHEUSSER (1972) indicated Fr_1 = 2.85 and 6.0. A re-analysis of their data suggests that, for partially-developed inflow conditions, Fr_1 = 2.98 and 8.04 respectively, and that Fr_1 = 3.26 and 7.32 for fully-developed hydraulic jumps.

Table 6-9 - Air bubble sizes in the turbulent shear region of hydraulic jumps

Ref.	Inflow conditions	V_1	Fr_1	C	$(d_{ab})_{max}$	$(d_{ab})_{mean}$
		m/s			m	m
(1)	(2)	(3)	(4)	(5)	(6)	(7)
THANDAVESWARA	P/D	2.48	7.16		0.0020	0.0012
(1974)	P/D	2.48	7.16		0.0043	
	P/D	2.66	7.41		0.0020	0.00115
	P/D	3.92	12.1		0.0006	0.00045
	P/D	4.22	12.5		0.0005	0.00038
	P/D	4.60	13.3		0.0005	0.0004
	P/D	3.99	10.3		0.0020	0.0011
	P/D	2.48	7.16		0.0043	
RESCH et al. (1974)	P/D	1.84	2.85	0.107	0.0300	0.0092
	F/D	2.01	2.85	0.051	0.0350	0.0104
BABB and AUS (1981)	P/D	3.52	6.00	0.10	0.0100	0.0060

Notes : F/D :   upstream flow with fully developed boundary layer

     P/D :   upstream flow with partially developed boundary layer

     $(d_{ab})_{max}$:   maximum bubble size in the turbulent shear region

     $(d_{ab})_{mean}$:   mean bubble size in the turbulent shear region

Air bubble sizes in the bubbly flow region

LEUTHEUSSER et al. (1973) recorded bubble size distributions using hot-film probes. They observed skewed distributions indicating a preponderance of bubbles of smaller size than the mean (fig. 6-6). Their data indicated also some very large bubbles as observed near the free-surface.

In the turbulent shear zone, the size of the bubbles is controlled by the turbulent breakup process (paragraph 6.2.3). Experimental data of maximum and mean bubble sizes in the turbulent shear region have been re-analysed (table 6-9). The results are shown in figure 6-7. Figure 6-7 shows that the maximum bubble size decreases with increasing upstream flow velocity. Indeed the upstream flow velocity is a measure of the level of turbulence in the shear layers of the jump. And the maximum bubble diameter is expected to decrease with increasing turbulence.

For upstream velocities between 1.5 and 5 m/s, the maximum bubble size and the mean bubble size in the shear flow region of a hydraulic jump can be correlated by :

$$(d_{ab})_{max} = 0.230 * V_1^{-3.93} \tag{6-37}$$

$$(d_{ab})_{mean} = 0.051 * V_1^{-3.08} \tag{6-38}$$

where V_1 is the upstream velocity in metres per second and d_{ab} is in metres.

Fig. 6-7 - Maximum and mean air bubble sizes in the shear flow region of hydraulic jumps - THANDAVESWARA (1974), RESCH et al. (1974), BABB and AUS (1981)

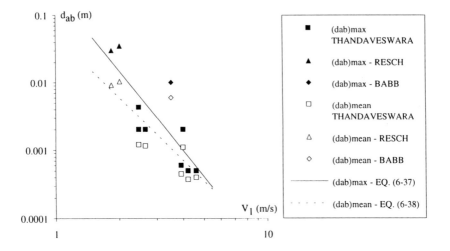

Air-water gas transfer

At a hydraulic jump, the major contribution to air-water gas transfer results from the small air bubbles entrained within the turbulent shear region (fig. 6-5). The turbulent shear region is characterised large air contents, small bubble sizes and hence large interface area. In the shear region, the air-water interface area is of the order of magnitude of :

$$a \sim 6 * \frac{C_{max}}{(d_{ab})_{max}} \qquad\qquad (6\text{-}39)$$

where C_{max} is the maximum air content in the turbulent shear region and $(d_{ab})_{max}$ is the maximum bubble size (eq. (6-37)). For hydraulic jumps with partially-developed inflow, CHANSON and QIAO (1994) showed that the maximum air content in the turbulent shear region is observed next to the jump toe and is estimated as :

$$C_{max} = 0.1467 * (V_1 - 0.414) \qquad\qquad (6\text{-}40)$$

where V_1 is in metres per second.

For a constant temperature and assuming a quasi-constant interface area (eq. (6-39)), the air-water gas transfer equation (eq. (6-4)) can be integrated over the jump. It yields :

$$r \sim \exp(K_L * a * t) \qquad\qquad (6\text{-}41)$$

where t is the residence time of the air bubbles. To a first approximation, the residence time t can be approximated as :

$$t \sim 2 * \frac{L_a}{V_1} \qquad\qquad (6\text{-}42)$$

where L_a is the aeration length (eq. (3-28)). Equation (6-42) assumes that the air bubbles are entrained with a mean velocity equal to $V_1/2$ over the aeration region.

With these assumptions (eq. (6-39) and (6-42)), an estimate of the air-water gas transfer at a hydraulic jump is :

$$r \sim \exp\left(12 * K_L * \frac{C_{max}}{(d_{ab})_{max}} * \frac{L_a}{V_1} \right) \qquad\qquad (6\text{-}43a)$$

Note that C_{max}, $(d_{ab})_{max}$, and L_a are functions of the inflow conditions of the hydraulic jump only. The fluid properties are taken into account by the coefficient of mass transfer. Replacing by equations (3-28), (6-37) and (6-40), it yields :

$$r \sim \exp\left(13.39 * \frac{K_L}{g} * V_1^{4.93} * (V_1 - 0.414) * \frac{\left(\sqrt{1 + 8*Fr_1^2} - 1\right)*(Fr_1 - 1.5)}{Fr_1^2} \right) \qquad (6\text{-}43b)$$

Equations (6-41) and (6-43) enable to estimate the gas transfer at a hydraulic jump *with partially developed inflow*. These are based upon physical reasoning and take into account the fluid mechanics properties of the jump.

WILHELMS et al. (1981) injected concentrated solutions of Krypton-85 upstream of a hydraulic jump and they recorded the downstream gas concentration. The saturation concentration for Krypton-85 is zero and their experiments analysed basically the desorption of the gas. Equation

(6-43) can be applied to the experiments of WILHELMS et al. (1981). The mass transfer coefficient is estimated using KAWASE and MOO-YOUNG's (1992) correlation with a molecular diffusivity of Krypton-85 in water of 1.92E-9 m^2/s at 25 Celsius (HAYDUK and LAUDIE 1974). The results are presented on figure 6-8 where the aeration efficiency at 25 Celsius, computed with equation (6-43), is compared with the data adjusted to 25 Celsius (eq. (6-7)). Figure 6-8 shows a reasonable agreement between the data and computations. Some scatter is observed for the largest efficiencies and highest Froude numbers. But the differences might result from measurements errors : WILHELMS et al. (1981) stated that "the reaeration rate coefficients vary greatly for replicate tests" and suggested additional tests at high Froude numbers.

Further, figure 6-9 presents a comparison of oxygen transfer calculations between the correlations of AVERY and NOVAK (1978) and WILHELMS et al. (1981), and equation (6-43). The comparison is done within the range of the experimental studies (table 6-5). Substantial differences are noted for large Froude numbers. But the scatter is within the errors : both the data of AVERY and NOVAK (1978) and WILHELMS et al. (1981) exhibited large inaccuracies for the largest Froude numbers.

Equation (6-43) enables a prediction of the air-water gas transfer at *hydraulic jumps with partially developed inflow* of the same order of magnitude as experimental observations. As the gas transfer model is based upon physical evidence, it is believed that this new model could provide better gas transfer predictions on prototypes than empirical correlations based upon model data. Further, equation (6-43) can be applied to any dissolved gas and to both dissolution and desorption situations.

Discussion - Air-water gas transfer calculations

Various researchers measured distributions of air concentration and velocity in hydraulic jumps (e.g. table 6-8). But none of these studies performed simultaneous air concentration and bubble size measurements. No information is available on the distribution of air-water interface area.

With partially-developed hydraulic jumps, equation (6-43) enables to predict the air-water gas transfer. But additional experiments are required to confirm the validity of the model for large Froude numbers and on prototypes. For hydraulic jumps at deep plunge pools, the method of JOHNSON (1984) can provide a good estimate of the oxygen transfer. In other hydraulic jump situations, the empirical correlations of AVERY and NOVAK (1975) and WILHELMS et al. (1981) can be used to approximate the rate of transfer of dissolved oxygen (paragraph 6.2.2.3). But great care must be taken if these formula are used outside of the range of validity (table 6-5).

Fig. 6-8 - Air-water transfer of Krypton-85 at hydraulic jumps in term of aeration efficiency E at
25 Celsius - Comparison between the data of WILHELMS et al. (1981) and equation (6-43)

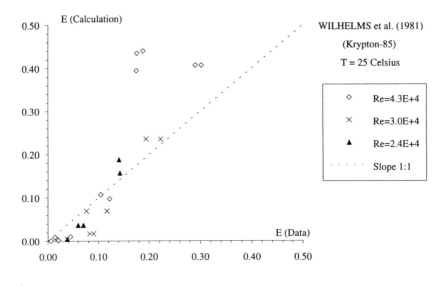

Fig. 6-9 - Oxygen transfer at hydraulic jumps : aeration efficiency E at 15 Celsius as a function of
the upstream Froude number Fr_1 - Comparison between equation (6-43) and the correlations of
AVERY and NOVAK (1978) and WILHELMS et al. (1981)

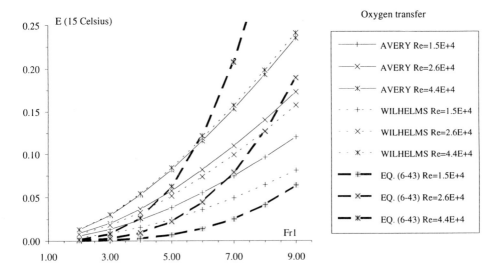

6.3.4 Discussion : gas transfer in nappe flow

The above paragraphs have shown the complexity for predicting the air-water gas transfer in a nappe flow regime. The total gas transfer results from the successive aeration of the flow at each step (paragraph 6.3.1). At each individual step, the flow is aerated by a combination of air bubble entrainment at the intersection of the falling nappe with the receiving pool of water on the step, and air entrainment at the following hydraulic jump.

The mechanisms of air entrainment by plunging jet and by hydraulic jump are complex. Although experimental investigations are under way in several laboratories, the author believes that accurate prediction methods will not be available before few years.

Design considerations : oxygen transfer

For design purposes, a crude estimate of the dissolved oxygen transfer can be obtained by "feeding" empirical correlations (e.g. AVERY and NOVAK 1978) into equations (6-28) or (6-29). The results can be double-checked with overall correlations (e.g. Department of Environment 1973, eq. (6-9)). Such approximations might provide an order of magnitude of the downstream DO content if the flow conditions are within the range of validity of the empirical correlations; but they will become very inaccurate outside of the range of validity of the empirical correlations.

At each individual step, the aeration efficiency of the plunging jet and the hydraulic jump can be estimated using the nappe flow properties (eq. (3-6) to (3-12)) and the correlations of AVERY and NOVAK (1978) :

$$(E^{jet})_{15} = 1 - \left(1 + 0.9525 * k_{AN} * \left(g * \frac{h^3}{v_w^2}\right)^{0.265} * \left(\frac{d_c}{h}\right)^{-0.534}\right)^{-1} \quad \text{for a wide channel} \quad (6-44)$$

$$(E^{HJ})_{15} = 1 - \left(1 + 6.965 * k' * \left(g * \frac{h^3}{v_w^2}\right)^{0.375} * \left(\frac{d_c}{h}\right)^{0.259}\right)^{-1} \quad (6-45)$$

where k_{AN} and k' are given in tables 6-4 and 6-6 respectively. For a channel with N_{step} equal steps, the oxygen transfer in term of aeration efficiency can be deduced from equations (6-44) and (6-45) :

$$E_{15} = 1 - \left(1 + 0.9525 * k_{AN} * \left(g * \frac{h^3}{v_w^2}\right)^{0.265} * \left(\frac{d_c}{h}\right)^{-0.534}\right)^{-N_{step}}$$

$$* \left(1 + 6.965 * k' * \left(g * \frac{h^3}{v_w^2}\right)^{0.375} * \left(\frac{d_c}{h}\right)^{0.259}\right)^{-N_{step}} \quad (6-46)$$

For nappe flows with fully-developed hydraulic jumps, equation (6-46) is compared with experimental data (ESSERY et al. 1978) in figure 6-10. The results suggests that equation (6-46) overestimates the gas transfer. Indeed, on stepped channels, the gas transfer by plunging jet is limited by the thin water depth on the receiving step while equations (6-44) and (6-46) are based on overfall data with deep receiving pool of water. The author believes that equation (6-44) overpredicts the gas transfer by plunging jet in nappe flow situations.

Fig. 6-10 - Oxygen transfer in nappe flows with fully developed hydraulic jump at 15 Celsius - Comparison between experimental data (ESSERY et al. 1978) and computations (eq. (6-46))

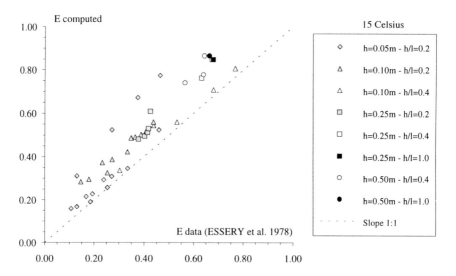

Discussion

For deep pooled steps, most air entrainment and air-water gas transfer result from the plunging jet process. Although the complete integration of equation (6-18) is not possible because of the lack of information on the velocity distributions, empirical correlations (e.g. HOLLER 1971, AVERY and NOVAK 1978, NAKASONE 1987) might provide some estimate of the air-water transfer of oxygen.

For nappe flows with shallow water depths, most air entrainment by plunging jet takes place at the inner shear layer of the jet (fig. 6-1 and 6-2). The shallow waters prevent the downward diffusion of bubbles and the residence time of the bubbles is very small. The author believes that, for nappe flow regime with small discharges, the contribution of the plunging jet to the total air-water gas transfer is small and nearly negligible. Most of the gas transfer occurs at the hydraulic jumps.

6.4 Skimming flow regime

In a previous part (chapter 4), similarities between free-surface aerated flows on smooth chutes and stepped chute flows with skimming flow regime have been shown. Using this analogy, the air-water gas transfer of skimming flows on stepped chutes can be estimated using the same method as CHANSON (1993d). Such a method requires an estimate of the distributions of undissolved air content and of bubble sizes.

6.4.1 Air bubble diffusion

In self-aerated flows on smooth chutes, the air concentration distribution can be represented by a diffusion model of the air bubbles within the air-water mixture (WOOD 1984) that gives the shape of the air concentration distribution for all mean air concentrations :

$$C = \frac{B'}{B' + \exp\left(-G'^{*}\cos\alpha^{*}\left(\frac{y}{Y_{90}}\right)^{2}\right)} \tag{6-47}$$

where C is the local air concentration defined as the volume of undissolved air per unit volume of air and water, B' and G' are functions of the mean air concentration only (appendix D), α is the spillway slope, y is the distance measured perpendicular to the invert and Y_{90} is the depth where the air concentration is 90%. The values of B' and G' are given in appendix D.

Next to the invert, prototype and model data (CHANSON 1992b) depart from equation (6-47), and show consistently the presence of an air concentration boundary layer in which the air concentration distribution is :

$$\frac{C}{C_{b}} = \left(\frac{y}{\delta_{ab}}\right)^{0.270} \tag{6-48}$$

where C_{b} is the air concentration at the outer edge of the air concentration boundary layer and δ_{ab} is the air concentration boundary layer thickness. On smooth spillways, the author (CHANSON 1992b) estimated δ_{ab} = 15 mm. C_{b} satisfies the continuity between equations (6-47) and (6-48). For spillway flows, the characteristic depth Y_{90} is much larger than δ_{ab} and a reasonable approximation is : $C_{b} \sim B'/(1+B')$.

Although equations (6-47) and (6-48) were obtained for smooth chute flows, it is believed that they might provide a reasonable estimate of the air bubble diffusion in self-aerated flows above stepped chutes.

6.4.2 Air-water interface area

The maximum bubble size can be computed combining the 1/N-th power law velocity distribution (eq. (4-40)) and equation (6-23). It yields :

$$d_m \sim \sqrt[3]{2 * N^2 * \frac{(We)_c}{(We)_e} * \left(\frac{y}{Y_{90}}\right)^{2*(N-1)/N}}$$
(6-49)

where N is the velocity distribution exponent and $(We)_e$ is the self-aerated flow Weber number :

$$(We)_e = \rho_w * \frac{V_{90}^2 * Y_{90}}{\sigma}$$
(6-50)

where V_{90} is the characteristic velocity at $y = Y_{90}$. With the hypothesis that the critical Weber number $(We)_c$ equals unity and the bubble diameter is in order of magnitude of the mixing length, equation (6-49) provides an estimate of the maximum bubble size in self-aerated skimming flows.

High speed photographs showed that the shape of air bubbles in self-aerated flows is approximately spherical and the specific interface area equals :

$$a = 6 * \frac{C}{d_{ab}}$$
(6-51a)

for air bubbles in water. For water droplets in air, the interface area is :

$$a = 6 * \frac{1 - C}{d_{wp}}$$
(6-51b)

where d_{wp} the diameter of water particles.

The calculation of the air-water interface requires the definition an "ideal" air-water interface. An advantageous choice is the location where the air concentration is 50%. This definition is consistent with experimental observations (CHANSON 1993d) and satisfies the continuity of equation (6-51).

Assuming $\{d_{ab} \sim d_m\}$, equations (6-47), (6-48), (6-49), and (6-51) enable the calculation of the air-water interface area at any point as a function of the flow properties. On figure 6-11, equations (6-47), (4-40), (6-49) and (6-51) are plotted for a 45-degrees chute with $q_w = 2$ m^2/s and $C_{mean} = 0.47$. Such discharge and mean air content are close to that observed on Aviemore spillway (CAIN 1978).

Figure 6-11(A) show the typical distributions for a smooth chute, computed as CHANSON (1993d). Figure 6-11(B) shows the distributions of air concentration, velocity, interface area for a stepped chute with the same discharge and mean air content as for figure 6-11(A). Note that the interface area is smaller than for the smooth chute case (fig. 6-11(A)).

6.4.3 Air-water gas transfer at stepped chutes

For stepped channel flows, the characteristics of the inception point of free-surface aeration can be computed using equations (4-11) and (4-12) (see chapter 4). Downstream of the inception point (fig. 4-1), the hydraulic flow characteristics can be calculated using the same method as that developed by CHANSON (1993d).

Fig. 6-11 - Distributions of air concentration (eq. (6-47)), velocity (eq. (4-40)), maximum bubble
size (eq. (6-49)) and air-water interface area (eq. (6-51)) in self-aerated flows

(A) self-aerated flow on smooth chute

$q_w = 2$ m^2/s, $C_{mean} = 0.47$, $Y_{90} = 0.22$ m, $N = 6.0$

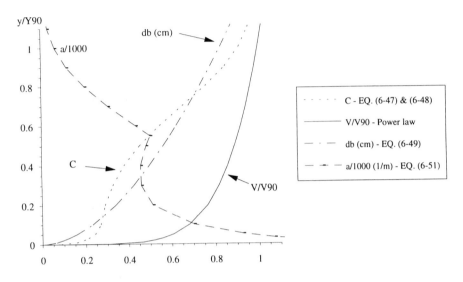

(B) skimming flow regime on stepped spillway

$q_w = 2$ m^2/s, $C_{mean} = 0.47$, $Y_{90} = 0.66$ m, $f = 1.0$, $h = 0.3$ m, $N = 1.13$ (eq. (4-39))

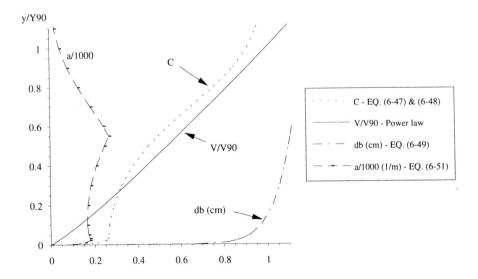

The aerated flow characteristics can be computed using a simple numerical model (CHANSON 1988, 1989b). The results validating the model for smooth spillway were presented elsewhere (CHANSON 1993a). The numerical model was modified for a stepped chute with skimming flow. The velocity distribution is estimated as equation (4-40) where the exponent N is deduced from equation (4-39). With this approximation, the air-water gas transfer due to self-aeration can be estimated at any position along a spillway (appendix D).

Typical results are shown on figure 6-12 where the aeration efficiency (for the dissolved oxygen and nitrogen contents) is plotted as a function of the distance from the spillway crest for several discharges and spillway slopes. Note that, as the water discharge increases, the inception point of free-surface aeration (eq. (4-11)) moves further downstream and the aeration efficiency due to free-surface aeration is reduced.

6.4.4 Discussion : self-aeration efficiency

The author performed a series of oxygen transfer calculations in self-aerated flows on stepped chutes, assuming an un-gated chute, zero salinity, constant channel slopes ranging from 15 up to 60 degrees, channel lengths between 20 and 250 metres, dimensionless discharges d_c/h from 0.8 to 21, a friction factor $f = 1.0$ and temperatures between 7 and 30 Celsius. The bubble rise velocity is assumed as that obtained on smooth spillway (i.e. $u_r = 0.4$ m/s, CHANSON 1993a).

The analysis of the oxygen and nitrogen transfer computations indicates that the free-surface aeration efficiency is independent of the initial gas content. Further the aeration efficiency (for oxygen and nitrogen) can be correlated by :

$$E = \left(1 - \frac{q_w}{(q_w)_c}\right)^{\Lambda} \qquad (6-52)$$

where $(q_w)_c$ is the discharge for which the growing boundary layer reaches the free surface at the spillway end and no self-aeration occurs, and the exponent Λ is a function of the dissolved gas, the temperature and the spillway slope.

For the range of the computations, the exponent Λ ranges from 3 up to 9 but no simple correlation has been obtained between Λ, the flow characteristics, the step geometry and the temperature. The characteristic discharge $(q_w)_c$ can be deduced from equation (4-11) :

$$(q_w)_c = 0.129 * (L_{spillway})^{1.403} * (\sin\alpha)^{0.388} * (h*\cos\alpha)^{0.0975} \qquad (6-53)$$

where $L_{spillway}$ is the spillway length.

On figure 6-13, equation (6-52) is plotted for $\Lambda = 3$ and 9, and is compared with skimming flow data obtained by ESSERY et al. (1978). Although the experiments of ESSERY et al. were affected by scale effects (section 6.1.3), their data show a trend similar to the air-water gas transfer computations (summarised by equation (6-52)). Free-surface-aeration calculations and experiments indicate that the aeration decreases when the discharge increases.

Fig. 6-12 - Aeration efficiency (dissolved oxygen and nitrogen contents) in a skimming flow
regime as a function of the distance along the spillway for several discharges - h = 0. 3 m, f = 1.0,
T = 293.15 K

(A) α = 15 degrees

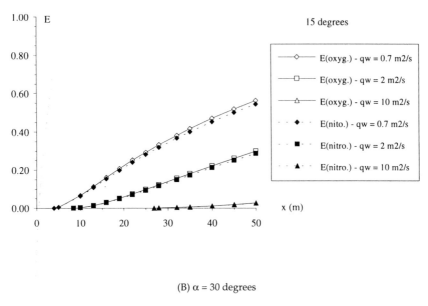

(B) α = 30 degrees

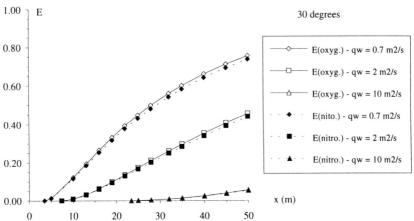

Fig. 6-12 - Aeration efficiency (dissolved oxygen and nitrogen contents) in a skimming flow regime as a function of the distance along the spillway for several discharges - h = 0. 3 m, f = 1.0, T = 293.15 K

(C) α = 45 degrees

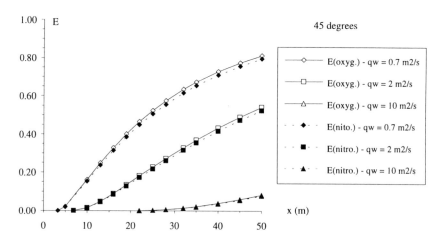

(D) α = 60 degrees

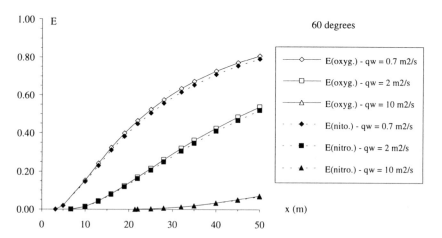

Fig. 6-13 - Aeration efficiency in skimming flow at 20 Celsius - Comparison between equation (6-52) and the skimming flow data of ESSERY et al. (1978)

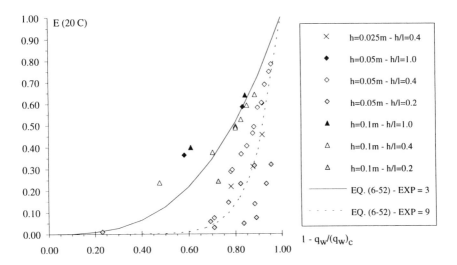

6.4.5 Discussion

Effect of channel slope

A comparison between the figures 6-12 suggests that the aeration efficiency due to free-surface aeration increases with increasing channel slopes between 15 and 45 degrees and is almost the same between 45 and 60 degrees.

For a given discharge, an increase in channel slope brings an increase of the amount of air entrained and hence and increase of the interface area. The mean velocity increases also with the channel slope and hence the residence time decreases with increasing slopes. For flat slopes, the quantity of air entrained is not large enough to obtain a optimum aeration. For steep channels, the mean flow velocity might become too large, the residence time becoming too short.

A channel slope of 45 to 50 degrees may provide a reasonable compromise for an optimum gas transfer efficiency.

Optimum air-water gas transfer in skimming flow

For a stepped chute with skimming flow, the above computations suggest that : 1- the air-water gas transfer due to self-aeration increases with decreasing water discharges for a constant channel slope, and 2- the aeration efficiency is maximum for channel slopes ranging from 45 to 60 degrees, for a given discharge.

A decrease in water discharge reduces the length of un-aerated flow. Hence the gas transfer due to free-surface aeration takes place over a longer distance. But the designer needs to keep in mind that the discharge must be large enough to satisfy the conditions of skimming flow regime (eq. (4-1), fig. 4-3).

Comparison with smooth chutes

The author (CHANSON 1993d) showed that free-surface aeration can contribute to a large part of the air-water transfer of atmospheric gases along a smooth chute. On a stepped chute, the air-water gas transfer along a stepped chute is further enhanced by : 1- the early apparition of "white waters" (i.e. free-surface aeration), 2- the slower mean velocity (and the associated increase of residence time), and possibly 3- the stronger turbulent mixing.

Figure 6-14 compares the aeration efficiency between stepped and smooth chutes for the same slope and discharge. The air-water transfer of oxygen is much larger on a stepped chute with skimming flow than on smooth chute.

The above method and computations (e.g. fig. 6-12) have taken into account only the early apparition of free-surface air entrainment and the slower flow motion. The author believes that the stronger turbulent mixing, observed on stepped spillway as compared to smooth chute, is expected to enhance the entrainment rate (i.e. $K_e > 1$ in CHANSON 1993a) and hence to increase the aeration efficiency.

The author recognises that the numerical model is complex. But equations (6-52) and (6-53), and figure 6-12, might provide an information on the order of magnitude for the air-water gas transfer (at 20 Celsius).

Fig. 6-14 - Comparison of air-water transfer of oxygen along smooth channel ($k_s' = 1$ mm) and stepped chute with skimming flow (h = 0.3 m, f = 1.0) - $q_w = 2$ m^2/s, $\alpha = 45$ deg., T = 293.15 K

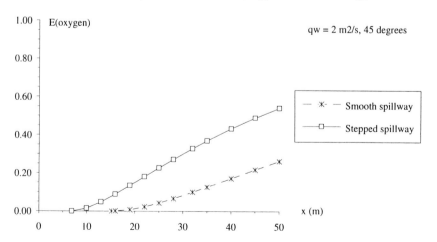

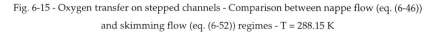

Fig. 6-15 - Oxygen transfer on stepped channels - Comparison between nappe flow (eq. (6-46)) and skimming flow (eq. (6-52)) regimes - T = 288.15 K

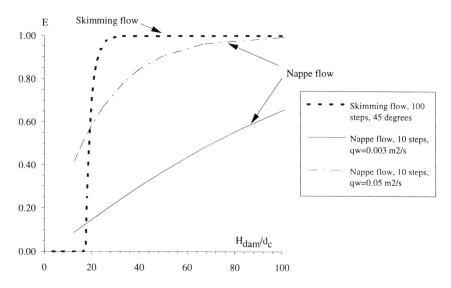

6.5 Discussion : comparison between nappe and skimming flows

Although the flow characteristics of nappe flow and skimming flow regimes are very different, a comparative analysis of the overall efficiency of each flow regime provides some interesting results. The author has investigated the effects of channel slope, discharge, step height and number of steps on the oxygen transfer.

The results indicate that the aeration efficiency of *nappe flows* is primarily a function of the discharge and channel length (fig. 6-15). The oxygen transfer increases with increasing channel length for a given discharge. Also the oxygen transfer increases with the discharge for a constant ratio H_{dam}/d_c. For nappe flow with fully developed hydraulic jumps, the data of ESSERY et al. (1978) indicate the same trend.

In *skimming flow* situations, the aeration efficiency is nearly zero as long as no free-surface aeration occurs. When free-surface aeration take place on the stepped channel, the oxygen transfer increases sharply with the dam height (fig. 6-15). The largest aeration efficiencies are obtained with small discharges. The analysis shows also that the channel slope and number of steps have little effects on the aeration efficiency of skimming flows.

In summary, the air-water gas transfer is maximum for low discharges with skimming flows; and it is maximum for large discharges in nappe flow situations. The apparent contrast of the conclusions can be explained by the different mechanisms of air entrainment and air-water gas transfer between the two flow regimes.

6.6 Examples of application

6.6.1 Air-water gas transfer in nappe flow regime

Two examples of applications are developed :

Example No. 1 : For small discharges, the flow aeration resulting from the plunging nappe is limited by the thin tailwater depth and the hydraulic jump is fully developed. Most air-water gas transfer results from the action of the hydraulic jump.

Example No. 2 : For deep pooled steps, the air-water transfer of chemicals occurs across the air-water surface area of the bubbles entrained by plunging jet, at the intersection of the free-falling nappes with the pool of water. No hydraulic jump occurs.

Example No. 1 : Nappe flow with a small discharge

Consider a cascade of 10 steps. The step height is 0.2 m, the step length is 0.75 m. For a water discharge of 0.03 m^2/s, compute the overall aeration efficiency of the structure for the dissolved oxygen at 15 Celsius.

We can verify that the flow regime is a nappe flow regime with fully-developed hydraulic jump. At the impact of the free-falling nappe with the step, the thickness of the pool beneath the nappe is d_p = 7.5 cm (eq. (3-8)) and the flow depth upstream of the hydraulic jump is d_1 = 1.6 cm (eq. (3-6)). With such shallow waters, the effects of plunging jet aeration are small to negligible.

Table 6-9 -Application for h $\doteq$ 0.2 m, q_w = 0.03 m^2/s, H_{dam} = 2 m

Variable (1)	Value (2)	Unit (3)	Eq. (4)	Comments (5)
d_c/h	0.226			
$(d_c/h)_{char}$	0.492		(3-14)	Fully developed hydraulic jump.
d_i	0.015	m	(3-10)	Nappe thickness at the impact with the receiving pool.
d_1	0.016	m	(3-6)	Flow depth upstream of hydraulic jump
Fr_1	4.66			Froude number at start of hydraulic jump.
d_p	0.075	m	(3-8)	Flow depth in pool beneath the nappe.
E^{HJ}	0.0545			Correlation of AVERY and NOVAK (1978)
E_1	0.0545		(6-26)	Aeration efficiency at one individual step.
E	0.429		(6-29)	Aeration efficiency of the 10-step cascade.

We will assume that most of the oxygen transfer takes place at the hydraulic jumps. The upstream Froude and Reynolds numbers of the hydraulic jump are 4.7 and 3E+4 respectively. Such values are within the validity range of the correlation of AVERY and NOVAK (1978) (table 6-5). At each step, the aeration efficiency of the jump is 5.45%. The total drop is 2 m and equation (6-29) yields to an overall aeration efficiency of 0.43. As a verification, the correlation of the Department of Environment (1973) gives a 74% aeration efficiency.

Our conservative calculations (table 6-9) suggest that the aeration efficiency is of the order of magnitude of 43% for the dissolved oxygen content.

Example No. 2 : Deep pooled steps

Consider a stepped cascade with 7 drops of equal height h = 0.5 m. Each step is pooled and the pool depth (in absence of flow) is 1.5 m. Compute the aeration efficiency of the structure for q_w = 0.05 m^2/s at 15 Celsius.

We will assume a wide structure (i.e. d << W). In operation, the mean free-surface elevation above the pool bottom equals : 1.5 m + 3/2 * d_c. The total aeration efficiency is deduced by adding the participation of each plunge. There is no hydraulic jump contribution.

For an individual step, the correlation of AVERY and NOVAK (1978) gives an aeration efficiency of 22%. Note that the flow conditions are within the range of validity of the correlation (table 6-2). For the overall structure, the aeration efficiency in term of dissolved oxygen content is about 83% (eq. (6-29)).

At each step, the nappe thickness at the impact is about 1.6 cm and the angle of the nappe with the horizontal is 70 degrees. Equation (6-32) gives a penetration depth of 3.3 m that is larger than the free-surface elevation (1.6 m). Hence our calculations (table 6-10) might overestimate the "real" flow aeration.

As a check, the correlation of the Department of the Environment (1973) yields to an aeration efficiency of 71%.

Table 6-10 -Application for h = 0.5 m, q_w = 0.05 m^2/s, H_{dam} = 3.5 m

Variable (1)	Value (2)	Unit (3)	Eq. (4)	Comments (5)
d_c/h	0.127			
d_b	0.045	m	(3-1)	Flow depth at the brink of a step.
d_i	0.016	m	(3-10)	Nappe thickness at the impact with the receiving pool.
V_i	3.11	m/s	(3-11)	Impact velocity of the nappe.
θ	70.4	degrees	(3-12)	Jet angle of the impinging nappe.
D_p	3.35	m	(6-32)	Penetration depth of the bubbles.
d_p	1.6	m		Free-surface elevation above the pool floor.
E^{jet}	0.223			Correlation of AVERY and NOVAK (1978)
E_1	0.223		(6-27)	Aeration efficiency at one individual step.
E	0.828		(6-29)	Aeration efficiency of the 10-step cascade.

Discussion

The calculations have shown that the penetration depth of the air bubbles is larger than tailwater depth. In such conditions, the correlation of NAKASONE (1987) might provide a more accurate prediction as it does take into account the effect of the tailwater level.

6.6.2 Air-water gas transfer in skimming flow regime

Considering a stepped weir with a slope of 45 degrees, the weir height is 20 m and the operating discharge is 5 m^2/s. The step height is 0.5 m. If the upstream dissolved oxygen content is 1E-3 kg/m^3, compute the downstream dissolved oxygen content at 20 Celsius. The salinity of the water is 18 ppt.

The boundary layer calculations provide the flow characteristics and the position of the inception point of air entrainment. The main results are given in table 6-11.

In summary : the water flows as a skimming stream above a steep slope (α > 27 degrees). The length of the free-surface aerated flow region is about 12 m. On figure 6-12(C), the aeration efficiency, 12-m downstream of the inception point, is around 0.18 and 0.05 for q_w = 2 and 10 m^2/s respectively. Note that equation (6-52) would predict an aeration efficiency between 0.16 and 0.004 for Λ = 3 and Λ = 9 respectively.

Assuming an aeration efficiency of about 0.115, the downstream DO content is approximately 1.8 E-3 kg/m^3.

Note

The calculations of dissolved oxygen transfer can be compared with smooth chute calculations.

For wide rectangular channels of constant slopes, the author (CHANSON 1993d) showed that the free-surface aeration efficiency (for oxygen) is independent of the initial gas content and can be correlated by :

$$E(\text{oxygen}) = \left(1 - \frac{q_w}{(q_w)_c}\right)^{(15.38 - 0.0351 * T) * (\sin\alpha)^{-1/3.13}} \qquad (6\text{-}54)$$

where $(q_w)_c$ is the discharge for which the growing boundary layer reaches the free surface at the spillway end and no self-aeration occurs. The characteristic discharge $(q_w)_c$ can be deduced from WOOD's (1985) formula :

$$(q_w)_c = 0.0805 * (L_{\text{spillway}})^{1.403} * (\sin\alpha)^{0.388} * k_s^{0.0975} \qquad (6\text{-}55)$$

where L_{spillway} is the spillway length and k_s is the roughness height. Equation (6-54) is valid assuming zero salinity, constant channel slopes ranging from 15 up to 60 degrees, channel lengths between 20 and 250 metres, discharges from 0.5 to 50 m/s^2, roughness heights between 0.1 and 10 mm (e.g. concrete channels) and temperatures between 5 and 30 Celsius.

For the present application, the aeration efficiency on a smooth chute (α = 45 degrees, H_{dam} = 20 m, q_w = 5 m^2/s) would be zero as the boundary layer would reach the free-surface downstream of the spillway end : i.e. $q_w > (q_w)_c$ (eq. (6-55)).

Table 6-11 -Application for h = 0.5 m, q_w = 5 m^2/s, α = 45 degrees, H_{dam} = 20 m

Variable (1)	Value (2)	Unit (3)	Eq. (4)	Comments (5)
d_c/h	2.7			Dimensionless critical depth.
$(d_c)_{onset}/h$	0.59		(4-1)	Implies a skimming flow regime.
L_I	16.1	m	(4-11)	Location of the point of inception.
d_I	0.53	m	(4-12)	Flow depth at the point of inception.
C_s	8.17E-3	kg/m^3	App. D	Saturation concentration of oxygen.
$(q_w)_c$	11	m^2/s	(6-53)	Characteristic discharge.
E (oxygen)	0.1145		(6-52)	Deduced from figure 6-12(C).
$C_{D/S}$	1.8E-3	kg/m^3		Downstream dissolved oxygen content.

CHAPTER 7
DESIGN OF STEPPED CHANNELS AND CHUTES

7.1 Introduction

7.1.1 Presentation

The design of a hydraulic structure is a long process which involves a substantial number of external factors : economics, environment, geology, hydraulics, hydrology, politics, sociology, ...

For the design of a channel (e.g. chute, spillway, weir), the first questions to ask are :

1- What is the purpose and the duty of the structure ?

E.g. : energy dissipation, water treatment, flood release, water supply ...

2- What are the constraints ?

E.g. : geometry, hydrology, material availability, topography, geology, politics ...

When the decision to select a stepped channel is done, all the parameters must be considered to decide the preferential type of flow regime :

A- nappe flow regime, or

B- skimming flow regime,

for the design discharges and for unusual (emergency) flow conditions (in drought or/and flood periods).

With stepped channels, the flow conditions near the transition between nappe and skimming flow regime must be avoided (if possible). Transitory fluctuations between nappe and skimming flow regimes might induce improper or dangerous flow behaviours, and unnecessary vibrations of the structure. In some cases (e.g. side channel), an incoming/outcoming discharge or/and a modification of the channel width or slope might induce a change of flow regime and it must be taken into account at the early stage of the design. A practical application is discussed in paragraph 7.1.2.

In this chapter, the author describes new design procedures for various applications of stepped chutes. Specific comments are detailed in each case and some design procedures are proposed. This chapter does not intend to review all possible cases but the main applications.

7.1.2 Flow regime transition - Discussion

The previous chapters have been developed for prismatic channels with constant discharges and step geometry. And the flow regime has been considered either nappe or skimming flow. But in real applications, a change of discharge per unit width q_w or a change of step geometry (h, l) might induce a change of flow regime. Flow conditions near the transition nappe-skimming flow are characterised by hydrodynamic instabilities leading to large fluctuating hydrodynamic loads on the steps. Designers must avoid such conditions and consider additional hydraulic and structural tests if they cannot avoid a flow regime transition.

In paragraph 4.2, it has been shown that the transition between nappe flow and skimming flow regime is a function of the dimensionless critical depth d_c/h and the channel slope (i.e. h/l). A modification of the channel width, an incoming or outcoming discharge (e.g. side channel) or a change of channel slope will entrain a modification of the discharge per unit width q_w and critical depth d_c, and possibly a change from nappe-to-skimming flow or from skimming-to-nappe flow.

To prevent a flow regime transition, the designer might consider a variable step height h, a variable step length l, a variable channel width W or a combination of these methods along the stepped channel, if the discharge is constant. With variable discharge, a transitory regime might not be avoidable. A particular example (i.e. side channel) is presented below.

A particular example of flow regime transitions is the Bank of China cascade in Hong Kong (fig. 7-3(D) and 7-3(E)) : incoming and outcoming discharges in a network of channels with identical width and step geometry provide a display of nappe and skimming flows.

Application : side stepped channel at a spillway bottom

Considering a side stepped channel at the bottom of a spillway (fig. 7-1), the stepped channel collects the flow discharging over the spillway (or weir). The discharge per unit width in the side channel changes with the incoming discharge and the modification of the channel width. At the upstream end of the side channel, the flow is a nappe flow regime and it might become a skimming flow with increasing discharge per unit width.

For the example shown on figure 7-1, the total discharge and the channel width vary both spatially. In the stepped channel, the discharge per unit width q_w can be deduced from the continuity equation :

$$\frac{d\,q_w}{ds} = q_w{}' * \frac{\cos\alpha}{W} - \frac{q_w}{W} * \frac{d\,W}{ds} \tag{7-1}$$

where s is the direction along the channel, $q_w{}'$ is the discharge per unit width of the spillway (i.e. incoming discharge), α is the stepped channel slope and W is the stepped channel width.

For a spillway of constant slope α' and constant discharge per unit width $q_w{}'$, and for a stepped channel of constant slope α, the critical depth at each position s equals :

$$\frac{d_c}{h} = \left(\frac{Q_w{}'}{W_1 * \sqrt{g * h^3}} * \left(\frac{\dfrac{s * \cos\alpha}{W'}}{1 + \dfrac{W'}{W_1} * \dfrac{\tan\alpha}{\tan\alpha'} * \left(1 - \dfrac{s * \cos\alpha}{W'}\right)} \right) \right)^{2/3} \tag{7-2}$$

where $Q_w{}'$ is the total spillway discharge, W' is the horizontal spillway width, W_1 is the stepped channel width at the downstream end and h is the step height (fig. 7-1).

Comparison between equations (4-1) and (7-2) will indicate the type of flow regime along the stepped channel. If $d_c < (d_c)_{onset}$, the waters flow as a nappe flow regime.

Fig. 7-1 - Side stepped channel at a spillway bottom

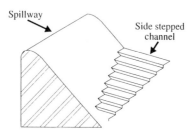

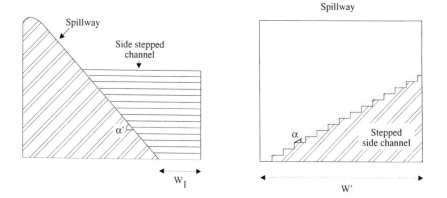

7.1.3 Comparison between flat and pooled steps

This monograph has analysed flat steps. The presence of a pool height at the step edges modifies substantially the flow pattern (i.e. streamlines and flow net) at the edge of the step. It is believed that the presence a pool height, even a very small (but finite) height, perturbs considerably the flow characteristics and that the results obtained with flat steps cannot be applied to a pooled step geometry. The existence of the pool height will affect the nappe trajectory in nappe flow regime, the onset of skimming flow and the flow resistance of skimming flows.

Application to nappe flow regime

In a nappe flow regime, the presence of a pool height at the step edge modifies the nappe contraction, the lower nappe free-surface and the free-fall trajectory. Figure 7-2 presents a comparison of the nappe trajectory downstream of a flat step, a sharp-crested weir and a finite pool height d_t. Further ROUSE (1938, fig. 163) showed photographic evidence that even a tiny height disturbs the streamlines.

Fig. 7-2 - Comparison of nappe trajectories between a flat step ①, a sharp-crested weir ③ and a
finite pool height ② - Ideal fluid flow solution (after ROUSE 1938)

Note : d_t = pool height

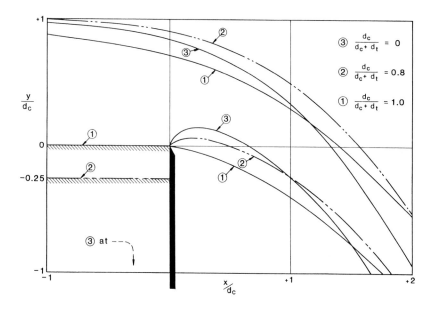

These examples highlight that the introduction of pool height is a major hydraulic decision.
Pooled-step channels cannot be designed using guidelines developed for flat-step chutes. With a
pooled-step geometry, designers are advised to conduct proper hydraulic investigations to
provide them with some guidance.

7.1.4 Spillway design

For the design of a spillway, the significant design parameters are the probable maximum flood
(PMF) discharge and the selection of the maximum discharge capacity. A recent investigation of
the causes of dam failures in the world since 1950 showed that more than 40% of the failures
were caused by overtopping (resulting from inappropriate spillway discharge capacity)
(LEMPERIERE 1993). These figures indicate clearly the importance and significance of an
appropriate estimation of the PMF discharge and a suitable selection of the maximum discharge
capacity.

In Europe, with temperate climates, the maximum spillway discharge capacity is much lower than the PMF discharge. But, in tropical countries (e.g. Northern Australia, Taiwan), the maximum discharge capacity is slightly smaller that the PMF. The standards depend also upon the risks of life danger.

Stepped spillways must be designed to sustain maximum design discharges without major damage to the spillway and to the dam itself. For embankment dams, the major flood risk relates to possible erosion of the dam, while the flood risk for concrete and masonry gravity dams concerns the stability rather than erosion. A gravity dam is designed for a given maximum headwater level and might overturn for an extra rise of headwater. A discussion of the risks of failures is further developed in chapter 8.

7.2 Design of stepped fountains

7.2.1 Introduction

In decorative architecture, waterfalls and cascades are used to please the eye and the ear. They provide a focal point as well as sounds generated by water splashing and cascading (fig. 7-3). Cascades can enhance the perception of a site by providing recreational facilities, an alteration of environmental factors to increase human comfort, a project image, a focal point or captivating views (AURAND 1986). In an oppressive environment (e.g. crowdy cities like Hong Kong and Tokyo), running waters can mask aggressive noise and help to provide an atmosphere of quietness and calm. Cascades and fountains serve as 'urban oasis'. Water can be also a significant element in interior landscapes where water provides simultaneously tranquillity and vitality.

Greek and Roman architects designed fountains and aqueducts several thousands years ago. Fountains and cascades have been used over the centuries as recreational and artistic features in most continents (table 7-1). Recently stepped cascades were combined also with leisure parks to enhance the water quality as well as providing recreational facilities (GASPAROTTO 1992).

7.2.2 Design procedure

Most of the attraction of stepped cascades is provided by the water splashing and the "white waters". Since the Antiquity, architects tried to estimate the most aesthetical fountain by trial-and-error, empirical correlations or personal experience.

Recent developments in stepped chutes and cascades (chapters 3 & 4) provide new information on their hydraulic characteristics, in particular on the apparition of "white waters". "White waters" result from the entrainment of multiple air bubbles by turbulent self-aeration at the free surface and by hydraulic jump.

Fig. 7-3 Examples of stepped cascades

(A) Chater Garden, Admiralty, Hong Kong in 1993

Thin free-falling nappes and intermediary pools

(B) Brisbane riverside, Australia in 1993 - Horizontal steps

Fig. 7-3 Examples of stepped cascades

(C) Keio Plaza Hotel, Tokyo, Japan in 1993 - Pooled steps

In designing a new cascade, the first step is selecting the flow regime : nappe flow or skimming flow. Special effects might be obtained also by combining the two flow regimes (e.g. Bank of China and Taipei World Trade Centre fountains).

In a *nappe flow* regime, most splashing occurs near the impact of the falling nappe on the step and at the downstream hydraulic jump. The main design parameters are : the location of the free-falling jet impact L_d, the length of the hydraulic jump L_r and the length of the bubbly region L_a (i.e. "white water" region) (fig. 3-4). These parameters are deduced as a function of the critical flow depth (i.e. $d_c = \sqrt[3]{q_w^2/g}$) and the step height h :

$$\frac{d_1}{h} = 0.54 * \left(\frac{d_c}{h}\right)^{1.275} \tag{3-6}$$

$$\frac{L_d}{h} = 4.30 * \left(\frac{d_c}{h}\right)^{0.81} \tag{3-9}$$

$$\frac{L_r}{d_1} = 8 * \left(\left(\frac{d_c}{d_1}\right)^{2/3} - 1.5\right) \tag{3-13}$$

$$\frac{L_a}{d_2} = 3.5 * \sqrt{\left(\frac{d_c}{d_1}\right)^{2/3} - 1.5} \qquad\qquad (3-28)$$

where d_1 and d_2 are respectively the flow depths upstream and downstream of the hydraulic jump.

Table 7-1 Characteristics of stepped cascades

Name (1)	Slope (deg.) (2)	Cascade height (m) (3)	Max. disch. (m^3/s) (4)	Step height (m) (5)	Nb of steps (6)	Type of steps (7)	Remarks (8)
Historical cascades							
Villa d'Este, Tivoli, Italy 1550						Flat steps.	Water staircases. Stepped channel [JE].
Chatsworth's Great Cascade, Derbyshire, UK 1696						Horizontal steps with intermediary pools.	Nappe flow regime [JE].
Peterhof's Grand Cascade, Leningrad, Russia 1715					7	Pooled steps.	Succession of waterfalls [JE].
Studley Royal, Fountains Abbey, UK 1725		3			5?		100-m long masonry dam [BI]. Nappe flow.
Public fountains							
Bank of China, Hong Kong (fig. 7-3(D) and 7-3(E))	17		0.002 to 0.022	0.04		Flat steps inclined downwards.	Nappe and skimming flows. Incoming & outcoming discharges. W = 0.25 m. l = 0.13 m.
Central Plaza, Hong Kong (fig. 7-3(F))						Flat steps, drops and pooled steps.	Nappe flow regime. Circular fountain.
Chater Garden, Hong Kong (fig. 7-3(A))				0.5 to 1	3	Pooled steps.	Square fountain. Thin free-falling nappes.
City riverside fountain, Brisbane, Australia (fig. 7-3(B))	23.2	3.45	0.011	0.15	6	Horizontal steps followed by a pool.	Nappe flow regime. W = 2 up to 3 m.
				0.15	17	Horizontal steps.	W = 5.5 up to 7.5(?) m.
The Forum, Central, Hong Kong					9	Pooled steps	
Peak tramway fountain, Hong Kong (fig. 1-2(I))	20	1.8		0.15	12	Horizontal steps.	Nappe flow regime.
Taipei world trade centre fountain, Taiwan (fig. 7-3(G) and 7-3(H))	~ 50				23	Horizontal steps.	Skimming then nappe flow regime. W = 0.8 to 5 m.
Keio Plaza Hotel, Tokyo, Japan (fig. 7-3(C))		2.4		0.4(?)	6	Pooled steps.	Succession of overfalls. No hydraulic jump.
Re-aeration cascades							
Calumet waterway system, USA 1991			1.52 & 0.91		3 & 4	Waterfalls.	5 artificial waterfalls [GA].

Note : [BI] BINNIE (1987); [GA] GASPAROTTO (1992); [JE] JELLICOE and JELLICOE (1971).

Fig. 7-3 Examples of stepped cascades

(D) Bank of China, Hong Kong in 1994 - Inclined downward steps

(E) Detail of the flow in the Bank of China fountain in 1994

Note the nappe flow in the left channel and skimming flow in the right channel

Fig. 7-3 Examples of stepped cascades

(F) Central Plaza, Hong Kong in 1994 - Combination of steps and drops

(G) Taipei World Trade Centre, Taiwan in 1994 - Horizontal steps

Fig. 7-3 Examples of stepped cascades

(H) Taipei World Trade Centre, Taiwan in 1994

Note the skimming flow at the upstream end and the nappe flow regime at the downstream end

In a *skimming flow* regime, the upstream flow is non-aerated and translucent. When the turbulence generated by the bottom boundary layer reaches the free-surface, air is continuously entrained through the air-water interface and the free-surface becomes "white". The distance L_I between the start of the stepped cascade and the apparition of the "white waters" is deduced from equation (4-11) :

$$\frac{L_I}{h} = 9.719 * (\sin\alpha)^{-0.277} * (\cos\alpha)^{-0.0695} * \left(\frac{d_c}{h}\right)^{1.0695} \qquad (7\text{-}3)$$

where α is the slope of the cascade ($\tan\alpha = h/l$).

7.2.3 Practical considerations

CROCKER (1987) indicated that waterfalls and cascades required a minimum discharge of about 0.006 m^2/s (i.e. $d_c > 15$ mm) to give a reasonable effect. In case of high wind speeds, he suggested to use a minimum discharge of 0.019 m^2/s (i.e. $d_c > 33$ mm). ERBE (1974) advised to use waterfall discharges less than 0.17 m^2/s to reduce the depth of the pools (i.e. d_c < 143 mm).

For stepped cascades, the first step must be carefully designed to avoid water jumping off the step. Designers may consider any combination of drops and steps (e.g. Central Plaza, Hong Kong, fig. 7-3(F)). They may choose variable channel widths and/or water discharges to display both nappe and skimming flows : e.g., variation of discharge at the Bank of China cascade (fig. 7-3(D)), change of channel width at the Taipei World Trade Centre fountain (fig. 7-3(G)) (see also paragraph 7.1.2).

With a thin overfalling nappe, nappe oscillations might develop and create air movements if the free-falling nappes are not adequately ventilated. In such cases, the sheet of water oscillates back and forth at a frequency of several hertz. The danger to the structure is small but the associated noise might be deafening and is always unwelcome (CASPERSON 1993). If the nappe cannot be aerated, splitters can be used to control the oscillations of the nappe. A maximum spacing between splitters of about two-thirds of the fall height (i.e. 2/3*h) is advised. Further information is given in paragraph 3.4.2 and appendix A.

7.3 Concrete stepped spillways and weirs

7.3.1 Introduction

Stepped spillways can be used to assist in energy dissipation and help to reduce the size (and the cost) of the downstream stilling basin. Typical examples are reported in table 1-2. For energy dissipation purposes,, a comparison of the energy dissipation with nappe flow regime and skimming flow regime (chapter 5) has shown that :

(a) for long channels where the uniform flow conditions are reached, the largest energy dissipation is obtained with a skimming flow regime on a steep chute (sub-regime SK3);

(b) for short stepped chute, a nappe flow with fully-developed hydraulic jump (sub-regime NA1) enables more energy dissipation than a skimming flow regime.

Hence the length of the channel (and the dam height) will affect the preferential flow regime to maximise energy dissipation.

7.3.2 Design procedure

A typical design procedure includes the following steps :

1- Assessment of the design flood

2- Assessment of the chute geometry

The main characteristics are the channel slope, chute height and width of the spillway. The spillway slope might be a design variable in some cases. A steep slope can minimise the volume of an overflow structure but larger energy dissipation is obtained with flat slopes.

3- Selection of the optimum step height

The step height is chosen as a function of the discharge and the channel slope to enable : the selected flow regime (nappe or skimming flow, paragraph 4.2) and the appropriate level of energy dissipation (chapters 3 and 5).

4- Selection of the step length

If the channel slope is not fixed by the topography or by the method of construction, the step length and the channel slope must be selected to provide the optimum flow conditions and/or energy dissipation characteristics.

5- Calculation of the hydraulic characteristics

For the selected step geometry, the flow depth, velocity, the amount of air entrainment and the energy dissipation are calculated using the information provided in chapter 3 (nappe flow regime) or chapters 4 & 5 (skimming flow regime). In particular, the effects of free-surface aeration must be taken into account for the design of the chute sidewalls and for the rate of energy dissipation.

6- Calculation of the downstream dissolved gas contents

For the climatic conditions of the site, and the chemical properties of the water, the aeration efficiency of the structure can be estimated (chapter 6). The most usual parameters are the dissolved oxygen and nitrogen gas contents (DOC and DNC). The prediction of the downstream dissolved gas contents must be performed for a wide range of discharges and atmospheric conditions.

Spillway crest

In the past, various shapes of spillway crest were used for the crest of stepped spillways. For skimming flows, model observations showed the risks of deflecting jets of water at the first steps if these are too high. Some researchers (SORENSEN 1985, BEITZ and LAWLESS 1992, BINDO et al. 1993) suggested to introduce few smaller steps near the crest to eliminate the risk of deflections.

The design of the first steps is still empirical and is deduced from model studies in most cases (e.g. fig. 1-1(C), fig. 4-1).

In Queensland (Australia), some stepped diversion weirs are designed with provision for an inflatable rubber dam. And negative pressures on the crest must be avoided since they produce destabilising conditions for inflatable dams. Such designs require usually a relatively long flat crest, inclined upwards (e.g. Bucca weir, fig. 1-2(C) and 7-5).

Nappe aeration

For a nappe flow regime, the ventilation of long free-falling nappes is essential to avoid unsteady fluctuations resulting from nappe oscillations (cf paragraph 3.4.2). This precaution is important for wide channels with large step heights at low flow rates.

In general, the most convenient method to ventilate the nappe is the use of sidewalls deflectors which provide space between the sidewalls and the deflected nappe. The design of these deflectors is somewhat similar to the design of aeration devices (CHANSON 1989a, WOOD 1991). It is possible also to aerate the nappe by small ducts feeding air from beyond the sidewalls to near the top of each drop. If these techniques are not feasible, splitters may be used to break the sheet of free-falling water, allowing space for aeration and preventing nappe fluctuations (fig. 3-7, eq. (3-17)).

Remarks

In some cases, the selection of the step height and channel slope might be determined by external factors rather than hydraulic considerations.

Step heights larger than 2-m might be selected to prevent people climbing up-and-down on stepped chutes. Steep slopes (i.e. $\alpha > 45$ degrees) or large step heights (i.e. $h > 2$ m) might be considered also to prevent motorcycles driving up and down the steps, as observed at the Brushes Clough reservoir (BAKER and GARDINER 1994) (fig. 7-13).   ·

In one particular instance, one spillway step of a diversion weir ($h \sim 2$ m, $l \sim 6$ m) was designed as a roadway for local farmers. It is used regularly during the dry periods. Such a 'roadway' step must be wide enough ($l > 2$ m) and adequately reinforced.

7.3.3 Roller compacted concrete (RCC) structure

Presentation

Roller compacted concrete (RCC) is defined as a no-slump consistency concrete that is placed in horizontal lifts and compacted by vibratory rollers.

Since the late 70's, RCC has become a popular material for the construction of gravity dams (table 1-2). The primary advantages of RCC gravity dams are : 1- the cost effectiveness, 2- the reduced foundation surface area compared to earthfill or rockfill structures of same height, and 3- the short time of construction (BASS 1993).

The low cost of RCC gravity dams results from : A- the small material volumes (using vertical upstream face and steep downstream face) compared with earthfill or rockfill structures, B- the lower cost per unit volume than conventional concrete, C- the construction techniques (little formwork, placement with standard earthwork techniques) and D- the reduced cost of auxiliary structures (intakes attached to the vertical face, stepped spillway on the downstream face) in comparison with rockfill and earthfill embankments.

Another application of RCC is the protection of earthfill embankments. In USA, the downstream slope of several small earth dams was protected by RCC overlays forming a stepped geometry. The downstream RCC slope enabled overtopping discharges. McLEAN and HANSEN (1993) reviewed over thirty projects.

Fig. 7-4 - Details of construction of steps on RCC dams

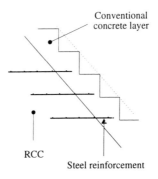

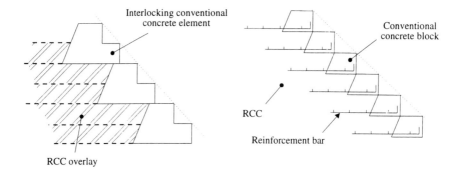

Discussion

With roller compacted concrete weirs, it is usual to shape the steps with a protective layer of medium to high-resistance concrete (fig. 7-4). Typical examples of construction techniques are summarised in table 7-2. During the construction, the joints between the RCC and the conventional concrete must be made with great care. Indeed model tests (FRIZZELL and MEFFORD 1991, LEJEUNE and LEJEUNE 1994) showed that the vertical face of the steps can be subjected to sub-atmospheric pressures. For stepped overflows above a RCC structure, it is essential to incorporate into the design a drainage system behind the erosion-resistant layer.

Figures 7-5 and 7-6 show construction details of a RCC diversion dam (Bucca weir, Australia). The dam was built between June 1986 and May 1987. It was made of 0.3-m thick RCC overlays with conventional concrete blocks protecting the downstream face. Figure 1-2(C) shows the dam after completion.

For small discharges and temperate climate, the steps might be simply formed RCC. McLEAN and HANSEN (1993) reported several cases of overflow over unprotected RCC faces (table 7-2). The stepped channels (with relatively flat slopes) were subjected to discharges up to 10 m²/s for several hours without damage despite the absence of protection layer.

In colder climates, unprotected RCC surfaces can experience some fracturing or chipping during freeze-and-thaw cycles. Some binders enable to entrain air in roller compacted concretes and 1.5-to-2% air content is sufficient to protect concrete from frost deterioration (DELAGRAVE et al. 1994). Alternatively, a conventional concrete overlay can be applied at the end or after the completion of the construction to protect the RCC.

Table 7-2 - Examples of stepped spillway construction for RCC dams

Dam	Ref.	Slope deg.	Step height m	Revetment layer	Comments
(1)	(2)	(3)	(4)	(5)	(6)
RCC dam with protective layer					
De Mist Kraal, South Africa 1986	[HD]	59	1	Conventional concrete.	RCC dam (117 kg cement/m^3). RCC overlay: 0.25 m thick.
Bucca weir, Australia 1987	[QW]	63.4	0.6	Conventional concrete placed before RCC and drilled by drainage holes.	RCC weir (90 kg cement & flyash/m^3). RCC overlay: 0.3 m thick. σ_c (90 days) = 20 MPa.
Les Olivettes, France 1987	[B]	53.1	0.6	Concrete (350 kg cement/m^3) with steel anchorage.	RCC dam (130 kg cement/m^3). RCC overlay: 0.3 m.
Riou, France 1990	[B]	59	0.6	Precast concrete blocks.	RCC dam (120 kg cement/m^3). σ_c (90 days) = 6 MPa.
New Victoria dam, Australia 1993	[W]	72 & 51.3	0.6	Slip-formed interlocking facing elements made of conventional concrete.	RCC dam (225 kg cement & flyash/m^3). RCC overlay: 0.3 m thick. σ_c = 20 MPa.
RCC without protective layer					
Brownwood Country Club dam, USA 1984	[MH]	26.6	0.23	Unprotected RCC over existing embankment.	Embankment dam (H_{dam} = 6 m). 0.23-m thick RCC overlay (184 kg cement & flyash/m^3).
White Cloud dam, USA 1990	[MH]	21.8	0.15	Unprotected RCC over existing embankment.	Embankment dam (H_{dam} = 4.6 m). 0.15-m thick RCC overlay (148 kg cement/m^3)
Camp Dyer diversion dam, USA 1992	[DF]			Unprotected RCC overtopping protection and RCC buttresses over existing masonry gravity dam.	H_{dam} = 21 m. Air-entrained RCC (134 kg cement & pozzolan/m^3, 2.5% air content). σ_c (90 days) = 13.8 MPa.
Horsethief dam, USA 1992	[MH]	26.6	0.30	Unprotected RCC over existing embankment.	Embankment dam (H_{dam} = 19.8 m). 0.305-m thick RCC overlay (193 kg cement/m^3)

Notes :

H_{dam} dam or weir height (m)

σ_c compressive strength (Pa)

[B] BOUYGUE et al. (1988); [DF] DOLEN and von FAY (1993); [HD] HOLLINGWORTH and DRUYTS (1986); [MH] McLEAN and HANSEN (1993); [QW] Queensland Water Resources (1988); [W] WARK et al. (1991,1992).

Fig. 7-5 - Cross-section of Bucca weir spillway (Australia)

Note the downstream stepped face (h = 0.6 m, h/l = 2) and the concrete protection layer

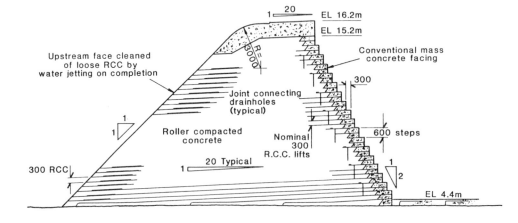

Fig. 7-6 - Construction of the Bucca weir on the Kolan river (Australia)

(Courtesy of Mr BEITZ, Queensland Water Resources Commission)

(A) 3 January 1987 - View from downstream

Fig. 7-6 - Construction of the Bucca weir on the Kolan river (Australia)

(B) 13 March 1987 - View from downstream - Note the completed portion of the crest on the left

(C) 9 April 1987 - Construction details of the crest viewed from the left bank

It is interesting to note that most masonry and concrete stepped spillways built before 1920 were reinforced with cut stones or granite blocks protecting the horizontal face of the steps (e.g. New Croton dam, Pedlar river dam) (see chapter 2). At that time, the strength of the concrete was much lower than today. Most concretes had a maximum allowable stress of about 0.5 to 1 MPa.

Cut stones and granite blocks were placed to protect the steps in a similar purpose as a conventional concrete protection layer on RCC dams.

7.4 Gabion stepped weirs and chutes

7.4.1 Introduction

Gabions are extensively used for earth retaining structures and for hydraulic structures (e.g. weirs, channel linings). Their advantages are : 1- their stability, 2- their low cost, 3- their flexibility and 4- their porosity. The porosity of gabions is an important characteristic, preventing the building-up of large uplift pressures. Typical design of gabion weirs are shown on figure 7-7. Some existing structures are presented in figures 7-8, 7-9, 7-10, 7-11 and 7-12.

Originally, a gabion[1] is a basket filled with earth or stone for use in fortification and engineering. The gabion technique of gabion has been known for thousands of years : original gabions used by Egyptians were made of rushes and papyrus. In China, hydraulic engineers used extensively gabions of bamboo basketwork[2] for over one thousand years.

Gabions used as construction material consist of rockfill material enlaced by a basket or a mesh. Various filling material and container can be used.

7.4.2 Description of the material

Box gabions are rectangular cages with rockfill material. Typical gabion dimensions are heights of 0.5 to 1 m, a width equal to the height and length-over-height ratio between 1.5 to 4. Long gabions are usually subdivided into cells by inserting diaphragms made of mesh panels to strengthen the gabion.

Wire and mesh

The wire is normally a soft steel with a zinc coating. In practice, the durability of gabion structures depends strongly upon the quality of the mesh and wires. The strength of the gabion container might be reduced by corrosion, gabion flexing and debris impacts. High quality and strengthened mesh wires can be used : e.g. galvanized wire, zinc coating wire, plastic (or PVC) coated wire.

Note that Chinese experience showed that bamboo baskets are durable and can last five to ten years at least.

---

[1]The word "gabion" ('gabion' in French, 'gabbione' in Italian) originates from the Italian 'gabbia' cage, meaning large cage ('cavea' in Latin).

[2]"Bamboo gabions" (or 'chu-lung') are made of woven bamboo slats packed with stones. Writings and paintings indicate that they were used before AD 1,000 as illustrated by NEEDHAM (1971).

Fig. 7-7 - Typical gabion weirs

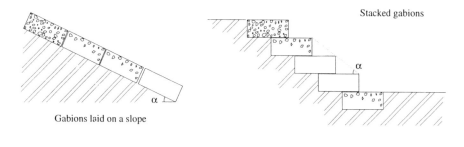

Stacked gabions

Gabions laid on a slope

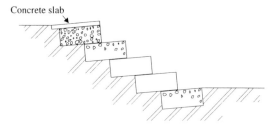

Concrete slab

Stacked gabions
inclined 5 degrees upward

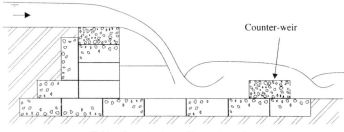

Counter-weir

Gabion drop structure with counter-weir

Filling material

The gabion filling consists of loose or compacted rocks. In any case, the stone size of the rockfill must equal at least 1 to 1.5 times the mesh size but should not be larger than 2/3 of the minimum dimension of the gabion. The use of small-sized stone (i.e. 1 to 1.5 times the mesh size) permits more economical filling of the cage and it allows a better adaptability of the gabions to deformation.

Fig. 7-8 - Gabion stepped weir in Yemen (1983) (Courtesy of Officine Maccaferri)

Note the concrete slabs on the steps to protect the gabions

Fig. 7-9 - Gabion stepped weir in Ontario, Canada (1980) (Courtesy of Officine Maccaferri)

Canalized water course near Highway No. 401

Fig. 7-10 - Gabion stepped spillway of the Guariraba weir at Campo Grande, Brazil (1984)

(Courtesy of Officine Maccaferri) - Note the seepage flow and the absence of overflow

The structural stability of gabions structures depends upon the weight of the gabions and the density of the filling material. The apparent unit mass of gabion is

$$[n_{gabion} * \rho_s] \qquad\qquad \text{(dry gabion)}$$

where n_{gabion} is the gabion porosity and ρ_s is the density of the filling material (table 7-3). The porosity of the gabion n_{gabion} is generally between 0.30 to 0.40. The density of gabion saturated with water equals

$$[n_{gabion}*\rho_s + (1-n_{gabion})*\rho_w] \qquad\qquad \text{(gabion saturated with water)}$$

Table 7-3 Density of different types of rocks (AGOSTINI et al. 1987)

Type of rock	Density ρ_s (kg/m^3)
(1)	(2)
Basalt	2900
Granite	2600
Hard limestone	2600
Trachytes	2500
Sandstone	2300
Soft limestone	2200
Tuff	1700

Table 7-4 - Stability criterion for gabion chutes

Reference (1)	Criterion of stability (2)	Comments (3)
Stacked gabions in steps		
STEPHENSON (1988)	$V < 4$ m/s	
PEYRAS et al. (1991)	$q_w < 3$ m^2/s	Gabions with strengthened mesh and lacing.
PEYRAS et al. (1992)	$q_w < 1.5$ m^2/s	Normal gabions.
Gabion laid parallel to the channel		
PEYRAS et al. (1991)	$q_w < 1$ m^2/s	Risk of overturning for larger discharges.

7.4.3 Risks of damage and destruction

Hydraulic erosion

The hydraulic performances of gabion stepped chutes are limited by the gabion resistance to abrasion and destruction, and their stability. Various researchers (table 7-4) proposed criteria to prevent the destruction of gabion spillways.

For gabions laid parallel to the flow, the stability of the gabion revetment depends upon the gabion dimensions but also upon the gap between gabions. PEYRAS et al. (1991) advised to design for discharges less than 1 m^2/s. But HUANXIONG and CAIYAN (1991) reported large overflow discharges (i.e. up to 34 m^2/s) on prototype structures without major damage. STEPHENSON (1979b, 1980) discussed the structural stability of stacked and aligned gabions.

For stacked (stepped) gabions, PEYRAS et al. (1992) recommended to use discharges less than 1.5 to 3 m2/s depending upon the mesh resistance. The step surfaces can be reinforced to enhance the abrasion resistance by timber, steel sheets, concrete facing (0.05 to 0.1 m thick) or even reinforced concrete slab (0.2 to 0.3 m thick) (AGOSTINI et al. 1987, PEYRAS et al. 1992) Figures 7-8 and 7-11 show some examples of concrete facing. With bamboo gabions, Chinese engineers use bamboo matting to protect the step surfaces.

Destruction of gabions

Stepped gabion weirs are subject to possible damage to the steps. Solid material carried by the stream may eventually abrade and fracture the gabion mesh. With large-size debris (e.g. trees), it is usual to protect the step surfaces with timber or concrete lining (fig. 7-8 and 7-11).

Gabion dams and gabion structures might be subjected to another risk of destruction. In some countries (e.g. parts of Africa or South-America), the inhabitants might use the gabion wires for other purposes (e.g. fences) or the gabion fillings as construction material. In such places, gabions should not be selected as construction material.

Discussion - Comparison with rockfill chutes

It is interesting to note that the discharge stability criteria for gabion spillways are of the same order of magnitude as the recommendations for unprotected rockfill channels (table 7-5).

Protected rockfilled slopes can achieve larger overflow discharges (table 7-5) but their cost is five times more expensive than unprotected rockfill (LEMPERIERE 1991). THOMAS (1976), ALLEN (1984) and LAWSON (1987) described several overtopping cases of rockfill dams. For short flood events (few days), discharges as large as 15 m^2/s were achieved over reinforced rockfill slopes without major damage.

Table 7-5 - Stability criterion for rockfill channels

Reference (1)	Criterion of stability (2)	Comments (3)
Unprotected rockfill OLIVIER (1967)	$q_w < 0.2335 * k_s^{1.5} * \left(\frac{\rho_s}{\rho_w} - 1 \right)^{1.667} * (\tan\alpha)^{-1.167}$	Model data (W = 0.56 m): $4.6 < \alpha < 11.3$ degrees $0.016 < k_s < 0.06$ m Gravel, pebbles, crushed granite (SI units).
KNAUSS (1979)	$q_w < \sqrt{g} * k_s^{1.5} * (2.7 - 3*\sin\alpha)$	Simplified formula based on model data (SI units).
ABT and JOHNSON (1991)	$q_w < 3.39 * k_s^{1.79} * (\tan\alpha)^{-0.768}$	Model data (W = 6.1 m) : $0.6 < \alpha < 11.3$ degrees $0.025 < k_s < 0.15$ m $\rho_s = 2650$ kg/m^3 (SI units)
LEMPERIERE (1991)	$q_w < 1$ m^2/s	Theories and models.
Rockfill protection LAWSON (1987)	$q_w < 15$ m^2/s	Mesh-protected rockfill dams.
LEMPERIERE (1991)	$q_w < 5$ to 10 m^2/s	Pervious downstream protection (reinforced rockfill).

Note : k_s median boulder size

ρ_s density of rockfill

W   channel width

Fig. 7-11 - Gabion river training, Loy Yang Mine, Victoria, Australia (Courtesy of Mr WILLIAMS, Geolab Group) - Note the partial concrete protection of the horizontal steps

Fig. 7-12 - Gabion torrent training in Saudi Arabia (1982) (Courtesy of Officine Maccaferri) Canalization of a torrent to protect a bridge on route 54 - Note the pooled steps

7.4.4 Design of gabion stepped chutes

The dimensions of the gabion and the design discharge determine normally the type of flow regime : nappe or skimming flow. Typically the height of the step h equals the height of the gabion. In some cases, h might equal twice or three times the gabion height. Usually the step length equals 1/2 to 1/4 of the gabion length. With these dimensions, figure 4-3 and equation (4-1) indicate the type of flow regime for the design discharge.

If the flow conditions are close the transition between nappe and skimming flow regime, the dimensions of the gabion and/or the width of the chute must be modified : either to change the step height h or to alter the discharge per unit width q_w.

For a structure made solely of gabions, the choice of a steep slope with a skimming flow regime would reduce the number of gabions and the cost of the structure. For a earth fill structure protected by gabions, a flat slope and a nappe flow regime may be more appropriate with the stability requirement of the embankment.

Inclined (upward) gabion-stepped spillways can also be used. Larger energy dissipation is achieved but their construction requires greater care. A recommended gabion tilt is : $\delta = +5$ degrees.

Discussion

With vertical drop structure, a counterweir is often used downstream of the primary structure to reduce the effect of erosion (fig. 3-3 and 7-7). The counterweir creates a natural stilling pool between the two structures which reduces the scouring force below the falling nappe impact.

The design considerations for the stability of gabion weirs are generally the same as for any gravity structure. The calculations of structural stability involve checking the stability of the weir against overturning, sliding and uplift.

Further seepage flow through the structure must be considered. The seepage flow is influenced by the infiltration caused by the spillway flow as well as by the flow through the gabion structure. CURTIS and LAWSON (1967) and KELLS (1993) discussed the associated problems. Figure 7-10 shows the seepage flow through a gabion weir in absence of overflow. Note the seepage flow increasing from the weir crest down to the downstream end.

7.5 Earth dam spillway with precast concrete blocks

7.5.1 Introduction

In recent years, several old embankment dams were identified as unable to pass the design flows without failure caused by overtopping. An additional discharge capacity can be obtained by designing a spillway made of concrete blocks on the downstream slope of the embankment (fig. 7-13).

Fig. 7-13 - Wedge concrete block spillway of the Brushes Clough dam, UK (1859/1991) (Courtesy of Mr GARDINER, NWW) - α = 18.4 deg., h = 0.19 m, inclined steps (δ = -5.6 deg.), trapezoidal cross-section (2-m bottom width, 1V:2H sideslope)

(A) Low flow spill in July 1993 (nappe flow regime)

(B) Details of the drainage holes (127mm×29mm)- Each concrete block weights about 120 kg

Soviet engineers were among the first to propose a stepped concrete protection on the downstream face of dam to pass flood discharges (table 1-2). The choice of a stepped structure allows the use of individual blocks interlocked with the next elements (PRAVDIVETS and BRAMLEY 1989) and assists in the energy dissipation. For new dams, a stepped spillway made of concrete blocks may be considered as the primary flood release structure of the embankment

(MILLER et al. 1987).

An interesting feature of such spillway is the flexibility of the stepped channel bed which allows differential settlements of the embankment. Individual blocks do not need to be connected to adjacent blocks.

7.5.2 Protective layer

For an earth dam with overflow spillway, the uppermost important criterion is the stability of the dam material. Seepage may occur in saturated embankment and the resulting uplift pressures might damage or destroy the stepped channel and the dam. An adequate drainage is essential.

For a typical design, the blocks lay on a filter and erosion protection layer (fig. 7-14). This layer has the functions to filter the seepage flow out of the subsoil and to protect the subsoil layer from erosion by flow in the drainage layer. Further the protection layer reduces or eliminates the uplift pressures acting on the concrete blocks.

Suction of the fluid from underneath the concrete steps can be produced by the pressure differential created by the high velocity flow over the step offset area (i.e. vertical face of the step) (FRIZELL 1992). Drains placed in areas of sub-atmospheric pressure will function to relieve uplift pressures (fig. 7-13(B)). In any case, the location of the embankment drains must be appropriately selected to avoid reverse flow in the drains and dynamic pressures associated with hydraulic jumps (in case of nappe flow regime). GRINCHUK et al. (1977) recommended that the total area of the drainage holes should be 10-15% of the block surface area.

FRIZELL (1992) studied flows past horizontal steps and steps sloped downward at 10 and 15 degrees. Her model tests suggested that the aspiration (on the vertical face of the steps) increased with the downward slope of the steps.

The seepage flow in the embankment dam must be predicted accurately to make the appropriate provision for drainage and evacuation of seepage flow though the blocks. Note that the seepage flow may be influenced by the infiltration into the downstream slope caused by the spillway flow, in addition to the flow through the embankment.

7.5.3 Block geometry and configuration

Soviet engineers developed a strong expertise in the design of concrete wedge blocks. PRAVDIVETS and BRAMLEY (1989) described several configurations (table 7-6). Blocks are normally laid parallel to the slope on the top of the protective layer (fig. 7-14). Individual blocks do not need to be connected to adjacent blocks. With such design, the channel bed is very flexible and allows differential settlements. For large discharges, the blocks may be tied to adjacent blocks and they would be made normally of reinforced concrete.

Fig. 7-14 - Details of concrete blocks on earth dam spillway

(A) flat concrete blocks (rectangular cross-section)

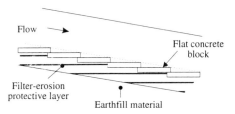

(B) flat concrete blocks of trapezoidal cross-section (non-overlapping block system)

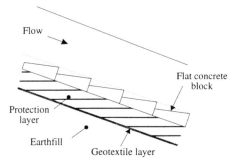

(C) wedge-shaped concrete blocks (overlapping block system)

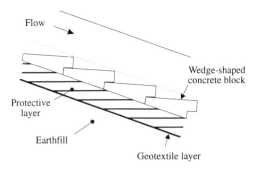

A related design option is the use of reinforced grass waterways (COLLETT 1975, HEWLETT et al. 1987). HEWLETT et al. (1987) tested a variety of reinforcement : textile, nylon, concrete blocks. Their results indicated that grass waterways, reinforced with concrete blocks, could sustain velocities up to 4 to 8 m/s without damage for a 3-hour test (table 7-6). Several grass species can be used : e.g. bent, couch, fescue, kikuyu, meadow grass, perennial ryegrass. In any case, COLLETT (1975) advised continual operation for no longer than 7 days, otherwise the grass can be adversely affected.

Table 7-6 - Design and stability of concrete blocks for embankment dam spillway

References	Max. disch. (m^2/s)	Slope degrees	Block shape	Block thickness m	Comments
(1)	(2)	(5)	(4)	(5)	(6)
GRINCHUK et al. (1977)	60 (V < 22 m/s)	8.7	Wedge-shape reinforced concrete block.	0.7	Dneiper hydro plant. Inclined steps downward.
PRAVDIVETS and BRAMLEY (1989)	2 to 27	7 to 15	Wedge-shaped block construction.	0.15 to 0.45	Inclined steps downward.
BAKER (1990)		21.8	Wedge-shaped blocks overlapping and non-overlapping (δ = - 8.3 deg.).		Model tests. No failure reported for : V ≤ 6.5 m/s.
Reinforced grass water way COLLETT (1975)	V < 2.2 to 2.7 m/s		Grass reinforced with wire		Reinforced grass water way.
HEWLETT et al. (1987)	V < 4 to 7 m/s	21.8	Interlocking non-tied concrete blocks.		Reinforced grass water way. 3-hour tests.
	V < 7 to 8 m/s	21.8	Interlocking tied concrete blocks.		

7.5.4 Design considerations

PRAVDIVETS and BRAMLEY (1989) recommended strongly that "the alignment of the spillway should be straight from the crest to the toe. Any curvature of the spillway in plan, or change in cross-section, will cause an uneven distribution of flows within the spillway which, in general, should be avoided".

Usually, the channel sidewalls are flat inclined slopes (i.e. trapezoidal spillway cross-section). The slopes of the sidewalls can be designed as inclined stepped surfaces (in the flow direction) and may use the same concrete block system as the main channel. PRAVDIVETS and BRAMLEY (1989) suggested typical sidewall slopes of about 1V:3H (i.e. 18 degrees).

7.6 Timber dams and crib dams

In countries and regions where transportation is difficult and timber plentiful, timber dams are serious competitors of concrete dams. In America, the North-East benefited from the experience of Northern European settlers and timber dams were reported as early as AD 1600. Prior to AD 1800, timber dams were usually 3 to 4.5-m high. At a later date, some much bigger ones were built successfully to a height of 30 m but most timber dams were less than 6-m high. Typical examples are detailed in table 1-2.

The life of a well-built timber weir can be estimated at 20 to 30 years. However dams reputed to be 80 to 100 year old are cited (ETCHEVERRY 1916, CREAGER et al. 1945). The construction cost of a timber dam is lower than a concrete structure. But the maintenance charges are substantial, particularly at sites where large floods and ice and debris runs are frequent. Leakage is frequent also and might be very large. A serious construction consideration is the tightening of the foundation.

ETCHEVERRY (1916) recommended to design timber dams with rockfill cribs and a stepped downstream face. He advised to use step length larger than 3 m. For $d_c/h < 4$, he suggested to use a downstream slope of $h/l = 1/3$, and $h/l = 1/6$ for $4 < d_c/h < 6$. At last, to facilitate the passage of ice and debris, he advised to use a curved crest with an inclined upstream face (slope of 1V:4H to 1V:2H).

A typical crib weir is shown on figure 7-15 : the Feather river dam. The Feather river weir was 5-m high and 85-m wide. The downstream face consisted of 4 steps (h = 2.4 m, l = 4.9 m). The structure was made of 0.09-m^2 square timber and the steps were covered by one layer of 0.15-m lumber. The design maximum discharge was about 73 m^2/s. One year after the construction, the weir discharged a flood of 54 m^2/s without damage.

Fig. 7-15 - Temporary timber dam on Feather river (USA, 1912) (after ETCHEVERRY 1916)

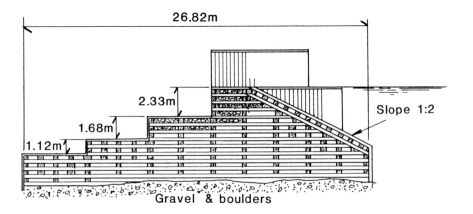

7.7 Debris dams

In mountain areas, where debris torrents might have catastrophic (and dramatic) impacts, "check dams" (or debris dams) can be used to reduce the impact of debris torrents. Debris dams are common features in Europe, North America and Far East Asia. Illustrations can be found in MAMAK (1964), VANDINE (1985), Soil and Water (1992) and on figures 7-16, 7-17, 7-18 and 7-19.

A debris dam can be made of woven fences, timber, earth, stone, riprap or concrete. *Timber debris dams* are cheap to build. The dam consists of a timber truss filled with stone. The crest can be paved or faced with logs. Typical life period of timber check dams range from 5 to 20 years. *Gabion check dams* provide reasonably-priced structures. The dam can be a simple vertical drop or a stepped weir. Culverts can be installed in the middle of the structure. Usually the dam crest is protected with steel or concrete facing. *Earth debris dams* are more appropriate for wide valley. The dam possesses a spillway section made of concrete or stones with sidewalls at least 3-m long. The upstream slope of the earth structure must be protected with riprap or pavement. *Stone dams* are more aesthetic. If the material is available in the vicinity of the dam location, the dam can be made of large boulder on cement. In cold regions, the quality of construction of the joints between boulders is very important : the action of the frost might damage (extensively) the masonry structure. *Concrete debris dams* are the most usual in developed countries. They are built as arch structures and are cheaper than stone structures.

Simple check dams have drop heights from 2 to 7 m. For higher dam height, it is advised to perform a stability analysis. Additional baffle blocks can be used to enhance energy dissipation. Trashracks can be installed upstream of the drops to stop coarse debris.

For concrete dam, MAMAK (1964) recommended the following geometry : the upstream face is nearly vertical, the downstream face has a 4V:1H slope, the crest width is about 2 m (fig. 7-16(A)). He suggested that the overflow discharge q_w should be less than 1.1 m^2/s for timber structure and less than 8.9 m^2/s for reinforced concrete.

Figure 7-16(B) shows a more sophisticated structure at Harvey Creek, Canada (HUNGR et al. 1987). The debris dam is an earth embankment with a double central drainage culvert and a overflow stepped debris spillway. The culvert is designed to carry the 200-year flood discharge. The upstream channel slope is covered with removable reinforced concrete beams. The beams can be lifted to inspect and clean the structure.

It is worth noting that a stepped bottom geometry is recommended also for avalanche corridors ("couloir d'avalanche") (JAMME 1974).

Fig. 7-16 - Examples of debris dams

(A) Typical European concrete debris dam

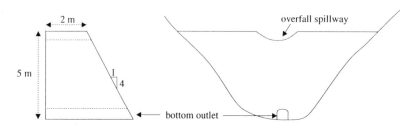

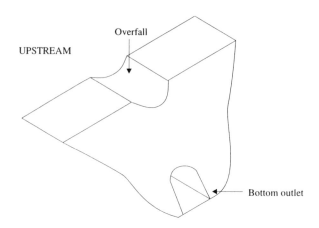

UPSTREAM

(B) Debris dam with culverts and stepped debris spillway

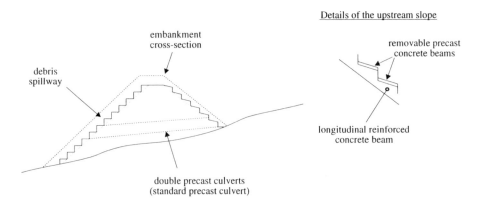

Fig. 7-17 - Check dams to control soil erosion during floods in Nigeria (1983) (Courtesy of Officine Maccaferri) - Note the openings for low flows

Fig. 7-18 - Soil retention check dam with a gabion stepped spillway (h = 1 m), Congohas, Brazil (1987) (Courtesy of Officine Maccaferri)

The colour of the flow indicates large soil-debris contents

Fig. 7-19 - Check dam with gabion stepped overflow near Sao Paulo, Brazil (1986) (Courtesy of Officine Maccaferri) - The dam is equipped with five culverts (0.6-m diameter pipes) to allow the passage of low flows and the gabion wires are PVC-coated because of the extensive air and water pollution of the area - H_{dam} = 11 m, h = 1 m

CHAPTER 8
ACCIDENTS, FAILURES AND SAFETY OF THE DESIGN

8.1 Presentation

Most hydraulic structures, including stepped channels and chutes, are designed for an optimum use at the most economical cost. The optimum design must minimize the 'real' total cost : i.e. the cost of construction + maintenance cost + cost of safety. The cost of a failure is not always measurable : damage to properties is assessable, damage to the environment is real and has often a political value, but loss of life is a matter of suffering.

In recent years, no hydraulic structure has failed because of a stepped channel geometry. But professional engineers must be aware of the possible consequences of a failure. In the past, a significant number of hydraulic structures failed, including stepped chutes. Some general reviews were prepared by ANCOLD, ICOLD, USCOLD and lessons can be learned from this experience.

In a first part, the author will review several accidents involving hydraulic structures with stepped channels and some related failures. Then considerations for a safe design of stepped channels will be developed. At last a discussion will highlight some important design aspects which must be considered by designers and operators of stepped chutes.

8.2 Failures and accidents

Over the past two hundred years, several accidents or failures occurred at hydraulic structures (e.g. dams, weirs) equipped with stepped channels. A summary is given in table 8-1. Each accident is investigated and discussed case by case.

In a second sub-section severe related accidents are described. The details of each accident are described and summarised also in table 8-1.

8.2.1 Failures and accidents of stepped channel structures

Arizona Canal dam (USA)

The Arizona Canal dam was located 45 km North of the city of Phoenix (USA). It was designed to divert water from the Salt river for irrigation purposes. Completed in 1887, the dam was a 280-m long structure made of timber cribs filled with loose rocks and gravels. The dam height was 10 metres and the base of the dam was about 11 to 14.6-m wide. At each end, the dam was fastened to bed rock. Beyond the rock, excavation was made in the river in which mud sills were laid upon which the cribs were built. Cribs were fastened together by wire cables.

The Western half of the dam was an overflow stepped spillway. It included 3 horizontal steps. At the time of construction, the maximum flood discharge observation was about 8,100 m^3/s (i.e. $q_w \sim 33$ m^2/s). It is believed that the overflow section was designed for that discharge.

In 1891, the central section of the dam settled considerably and portions of the dam were washed away. Prompt repairs were made. In 1899, signs of timber deterioration were observed in some upper portions of the dam. The dam was repaired again and strengthened.

On April 13, 1905, a major flood caused a dam break. At the time of failure, the discharge was about $q_w = 11.3$ m^2/s. Portions of the centre of the dam were torn away and the dam breach was about 91-m wide.

It is believed that the failure was caused by unstable foundation problems and some timber deterioration. Indeed, at the time of break, the dam was 18-year old. Further, at the time of failure, the ratio d_c/h was about 0.7 : i.e. the flow regime was near the transition between nappe flow and skimming flow (paragraph 4.2). The unstable flow situation might have induced additional fatigue to the structure.

Goulburn weir (Australia)

The Goulburn weir is located in North-Central Victoria (South-East of Australia). It is an important structure in the irrigation system of Northern Victoria, being the primary diversion of the Goulburn river to nearly 50,000 ha of irrigated land. Built in 1891, the weir is a concrete gravity dam (15-m high, 141-m wide). Flood waters discharge over the structure. The downstream face consists of stepped horizontal granite blocks (h = 0.5 m) (fig. 2-5). The maximum discharge capacity is about 1980 m^3/s. The structure was initially equipped with 21 cast and wrought gates (3-m high, 6-m wide) supported by cast iron piers.

In 1978, one gate failed to operate. The accident had no direct consequence but it initiated extensive inspections of the 90-year old structure. Inspections revealed serious deterioration to a significant number of gates and other damage in the recesses. Further the weir is founded on erodible weathered mudstone foundation. The inspections highlighted the need to stabilise the foundation and to prevent the risk of sliding.

Repairs were undertaken between 1980 and 1986. All the gates and piers were replaced by a system of 9 radial gates (3.65-m high, 12.87-m wide) supported by new concrete piers. The stability of the structure was secured by six prestressed anchor bars per pier and ground anchors at the spillway toe.

Lahontan dam (USA)

The Lahontan dam was built in 1915. The dam is an earth structure, 49-m high, across the Carson river (Nevada). The dam is owned by the US Bureau of Reclamation. The spillway consists of a succession of large steps. The maximum discharge capacity of the spillway is 742 m^3/s.

The concrete of the spillway was damaged by freeze and thawing (JANSEN 1983). No information is available on eventual repairs.

Minneapolis Mill dam (USA)

The Minneapolis Mill dam was built between 1893 and 1894. The structure was a masonry dam (5.5-m high, 163-m long). The crest was 1.6-m wide and the base of the dam was 3.66-m wide. The spillway was a 51.8-m long section of the dam, with 9 steps (h = 0.61 m, $\alpha \sim 75$ degrees).

In Minneapolis, the winter climate is severe and the ice of the reservoir adhered to the masonry structures during the winters. In February 1899, a slight motion of the dam was noticed. It was believed that the movement was caused by ice pressure. No cracks or leaks were discovered and steel anchors were placed to strengthened the structure.

On the 30 April 1899, the spillway was discharging 0.036 m^2/s when the anchors snapped and the weir slid out.

The failure was supposed to be initiated by cracks under the dam. These cracks were opened when the dam was slightly moved in February 1899 and uplift pressures caused the final collapse of the dam.

New Croton dam (USA)

The New Croton dam is a masonry dam built between 1892 and 1905. The dam was constructed across the Croton river, 65 km North of the city of New York. The structure is 90.5-m high and the base thickness is 63-m.

The spillway is an overflow stepped channel at the right-bank extremity of the dam (fig. 2-6). The spillway face consists of granite blocks with mortared joints. The spillway was designed for a maximum discharge of 1550 m^3/s (WEGMANN 1907). The spillway crest is 305-m wide and the slope is about 53 degrees. The step height is 2.13 m. Note that, during the construction, the spillway crest was raised by 1.2 m : 229 metres of the spillway crest were completed in masonry but 76 metres of crest close to the dam were obstructed by 1.2-m high flashboards and concrete slabs. Around 1907-1908, 0.6 metres of additional flashboards were added over the entire crest width to increase the reservoir capacity. The flashboards (and concrete slabs) were left permanently in place for 50 years.

Following a storm event on the 14 to 16 October 1955, the spillway was damaged by a water release of 651 m^3/s with an overflow depth of about 1.2 m. After the flood, inspections of the spillway revealed that the masonry close to the dam was badly cracked with several longitudinal cracks. The structure was no longer watertight and waters were escaping in significant amounts through the cracks. Extensive repairs were carried out between 1955 and 1957.

MARSH (1957) suggested that the masonry at the end of the spillway (close to the dam) was subjected to severe vibrations resulting from improper flow conditions below the added flashboards and concrete slabs. It is certain that, at the time of the flood, the spillway crest was incorrectly profiled. But the severity of the damage, for a discharge smaller than the maximum design discharge, suggests some concerns about the quality of the construction and the behaviour of the stepped spillway (GOUBET 1992).

Fig. 8-1 - Cross-section of the Puentes dam (after SCHUYLER 1909)
Note the wooden pile foundation (6.7-m deep) below the masonry structure

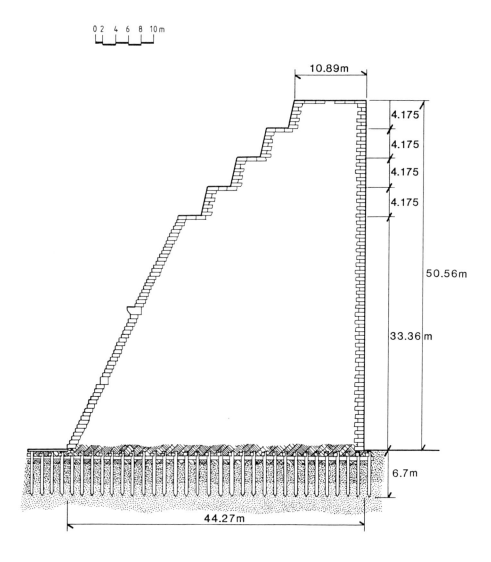

During the spill, the ratio of critical depth over step height (d_c/h) was 0.36. For a similar slope (51.3 degrees), BEITZ and LAWLESS (1992) observed the transition between nappe and skimming flow for $(d_c)_{onset}/h = 0.4$ (table 4-1). The comparison suggests that the flow regime during the spill of the New Croton dam was at the transition between nappe flow and skimming flow. The resulting flow instabilities might have added extra loads to the spillway structure, particularly below the flashboards. The fluctuating hydrodynamic loads might have accelerated the crack growth.

Puentes dam (Spain)

The Puentes dam was built between 1785 and 1791 on the Guadalantin river (Spain). The gravity dam was a rubble-masonry structure and the faces were made of cut stones. It was 50-m high, 282-m long, 11-m thick at the crest and 22.2-m thick at the base. The upstream face of the dam was vertical. The downstream face consisted of four steps (h = 4.2 m, l = 3.3 m) followed by an uniform slope down to the base (fig. 8-1). Flood waters were to be discharged over the dam crest (uncontrolled overflow). The dam was equipped with a large scouring gallery (6.7-m wide, 7.5-m high) at the base and two outlets (1.6-m wide, 1.95-m high each), one near the base and the second 30.5-m below the crest.

The masonry structure was founded on wooden piles driven into a bed of alluvial soil and sand. Hundred of wood piles (6.1-m long, 0.5-m diameter each) were driven into the alluvium and braced with horizontal timbers at their tops.

From 1791 to 1802, the reservoir was never filled and the depth of water behind the dam did not exceed 25 m. At the beginning of 1802, the water level began to rise and reached 46.8 m above the base of the dam in April. The spillway was not used before the dam collapse.

The failure occurred on the 30 April 1802. At the time of the accident, the water depth was 33.4 m and the mud deposited in the reservoir was 13.4-m deep. The water level was located below the dam crest. Although the dam in itself was sound enough, the upstream hydrostatic pressure blew a hole all the way underneath the dam. Immediately after, the central part of the dam collapsed and the stored water (~ 8 Mm^3) was quickly discharged. The reservoir was drained in less than one hour. The flood hit the downstream city of Lorca and six hundred and eight people were drowned.

After the rupture, the dam abutments were still standing on the hillsides and the dam breach was 17-m wide and 33-m high. The collapse was caused by the failure of the dam foundation.

St. Francis dam (USA)

The St. Francis dam, completed in 1926, was located in the San Francisquito Canyon, 72 km North of Los Angeles. The structure was a 62.5-m high curved concrete gravity dam. The dam thickness was 4.9 m at the crest and 53.4 m at the lowest base elevation.

The dam was equipped with 11 spillway openings (2.8 m^2 each) and 5 outlet pipes (Ø 0.7 m). The spillways discharged onto the downstream stepped face of the dam (h = 0.4 m). The overflow spillways were never used before the failure.

The reservoir began filling in 1926 and was nearly full on the 5 March 1928. The volume of the reservoir was 46.9 Mm^3 at that date. The dam site was investigated on the morning of the 12 March 1928 : nothing in the performance of the structure was judged 'hazardous' (OUTLAND

1963).

On the evening of the 12 March 1928, the collapse came all at once between 11:57 and 11:58 pm. The dam broke completely, releasing the 46.9 M^3 of water over 70 minutes. The peak discharge just below the dam reached 14,200 m^3/s. The resulting surge devastated bridges, highways, railroads, a construction camp and the town of Santa Paula (70 km downstream) before reaching the sea. The death toll was about 450 deaths. During the collapse, a single erect monolith survived essentially unmoved from its original position (fig. 8-2). After investigations, the remnant part was blasted by dynamite on the 23 May 1929.

The results of the investigation showed that the dam failure resulted from a combination of a massive landslide on the left abutment and uplift pressure effects. Afterwards, geological and geotechnical investigations revealed the very poor quality of the geological setting. JANSEN (1983) stated that "in view of the many deficiencies of the site, the survival of the structure for 2 years is remarkable indeed" !

Discussions (NOETZLI 1931, WILEY 1931) suggested positively the presence of large cracks in the interior of the gravity dam. Although the spillway was never used, the author wonders what could have been the effects of water release on the downstream stepped concrete face! Stepped spillway flows induce large hydrodynamic forces on the chute bottom and cavity recirculation creates fluctuating pressures on the step faces. The interactions between the large fluctuating efforts and the internal cracks could have been disastrous.

Fig. 8-2 - Catastrophe of the St. Francis dam, California (USA, 1928)

(A) View from downstream of the dam after the collapse

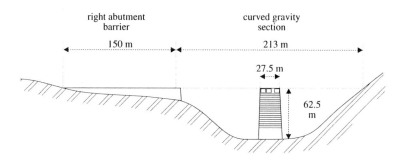

View from dowsntream of the St Francis dam
(after the collapse)

Fig. 8-2 - Catastrophe of the St. Francis dam, California (USA, 1928)

(B) View from downstream of the remnant part after the dam collapse

Warren dam (Australia)

The Warren dam is a concrete gravity dam initially built in 1916. The dam was raised in 1926 and again modified in 1974. The original dam was 17.4-m high and the spillway was a staircase channel (4 steps, h = 0.37 m). The maximum discharge capacity was approximately 100 m^3/s which corresponds to a once-in-forty-year flood.

In 1917, the structure was subjected to a record flood of 128 m^3/s during which the dam was overtopped. No failure or damage was observed. But the inadequacy of the spillway capacity was clearly highlighted.

In 1926, the structure was raised by 1.5 m and the spillway profile became an ogee weir followed by a smooth chute. The maximum discharge capacity remained 100 m^3/s. More recently, investigations of the safety of the dam suggested that the structure should pass a design flood of 177 m^3/s (1-in-100 year flood). In 1974, the spillway was modified again. The spillway crest was lowered by 1.07 m to allow the 177-m^3/s discharge without overtopping the dam.

Table 8-1 - Summary of accidents and failures

Dam/Reservoir	Ref.	Years of construction	Year of accident	Accident/Failure	Lives lost
(1)	(2)	(3)	(4)	(5)	(6)
Stepped spillways					
Puentes dam, Spain	[JA, SC, SM]	1785-1791	1802	Dam break (caused by a foundation failure).	608
Goulburn weir, Australia	[CU]	1891	1978	1- Gate failure (caused by corrosion) 2- Foundation stability	--
Minneapolis Mill dam, USA	[WE]	1893-1894	1899	Dam break during a small spill (caused by cracks resulting from ice pressure on the dam).	--
Arizona Canal dam	[EN]	1887	1891 and 1905	Partial destruction of the dam during a flood (caused by foundation problems and timber deterioration).	--
New Croton dam, USA	[FO, GO, MA, W7]	1892-1905	1955	Spillway damage during flood releases.	--
Lahontan dam, USA	[JA, RH]	1915		Damaged spillway concrete (caused by freezing and thawing).	--
Warren dam, Australia	[JO]	1916	1917	Dam overtopped (no damage).	--
St. Francis dam, USA	[JA]	1926	1928	Dam break (caused by foundation failures).	450
Smooth spillways					
Pinet dam, France	[LE, SO]	1927-1929	1982	Spillways blocked by debris.	--
Palagnedra dam, Switzerland	[BR, JA]	1952	1978	Dam overtopping (caused by a combination of large flood and large volume of debris).	24
Belci dam, Romania	[DI]	1958-1962	1991	Dam overtopping and breach (caused by a failure of gate mechanism).	97
Tous dam, Spain	[UT]	1977	1982	Dam break (following an overtopping; collapse caused by an electrical failure)	--

Note : [BR] BRUSCHIN et al. (1982); [CU] CUMMINS et al. (1985); [DI] DIACON et al. (1992);

[EN] Engineering News (1905a); [FO] FORD (1957); [GO] GOUBET (1992); [JA] JANSEN (1983); [JO] JOHNSON and TEMPLAR (1974); [LE] LEFRANC (1992); [MA] MARSH (1957); [SC] SCHUYLER (1909); [SM] SMITH (1971); [SO] SOYER (1992); [RH] RHONE (1990); [UT] UTRILLAS et al. (1992); [W7] WEGMANN (1907); [WE] WEGMANN (1911).

8.2.2 Related accidents

Recently, a number of hydraulic structures were subjected to severe accidents which might occur also with stepped channel structures. These involved obstructions of the spillway gates, failures of the gates and opening systems.

Belci dam (Romania)

The Belci dam was an earth embankment built between 1958 and 1962 across the Tazlau river, North-East of Romania. The dam was 18-m high with a concrete spillway (maximum discharge capacity : 1,400 m^3/s).

On the 28 and 29 July 1991, an exceptional flood reached the Belci reservoir. The peak inflow was 2,200 m^3/s. The dam was overtopped for almost 4 hours because of a failure to open and to operate the spillway gates. At the time of the overtopping, two out of the six spillway gates were partially opened and the others were closed.

A total of 98 people were killed or reported missing following the dam failure. Later investigations indicated that the accident was caused an incorrect prediction of the flood and a failure to open the spillway gates. The latter resulted from a succession of operator errors and power failures.

Palagnedra dam (Switzerland)

The Palagnedra dam is a 72-m high curved gravity dam. Completed in 1952, the original spillway included 13 ogee-crest openings (5-m wide, 3-m high) separated by piers supporting a road bridge. The design discharge was 450 m^3/s but later investigations showed that the maximum discharge capacity was larger.

In August 1978, an exceptional flood event occurred upstream of the reservoir and the flood waters washed away a great volume of debris. Some of these debris were blocked by a railway bridge, located upstream of the reservoir, and started to form a "debris" dam. When the bridge collapsed, a water-and-debris surge rushed to the reservoir and overtopped the dam (on the 7 August 1978).

Although the dam did not collapse, the structure was seriously damaged and had to be reinforced. Further twenty four (24) people perished in the event.

It was estimated that, during the flood, the maximum peak inflow to the reservoir reached around 3,000 m^3/s. The spillway discharged up to 1,000 m^3/s but was subsequently obstructed by debris. After the flood, the volume of debris blocked by the dam amounted to 25,000 m^3 of wooden debris and logs. Most debris were burned on the spot.

Pinet dam (France)

The Pinet dam is located on the Tarn river, 30 km downstream of the town of Millau. Built in 1929, the concrete gravity dam is 40-m high. The "old" spillway system included 18 radial gates (6.6-m wide each) and the design discharge was 3,500 m^3/s.

During the flood of November 1982, several spillway gates were obstructed by large debris (trees of 15 to 20-m length, camping cars). Following the incident, the public questioned the safety of the structure. The new hydrological investigations indicated that the flood discharge of the one-thousand-year return period event is 4,400 m^3/s. Further, the studies showed that the presence of debris reduced substantially the maximum discharge capacity. It was estimated that, in November 1992, the maximum discharge capacity was diminished down to 2,600 m^3/s.

Repair works were undertaken between 1987 and 1990. The 18-gate system was replaced by three 40-3-m long flap gates and the spillway crest was lowered by 1.1 m. As the result of these modifications, the maximum discharge capacity was increased to 4,400 m^3/s and the risk of gate obstruction was eliminated.

Tous dam (Spain)

Completed in 1977, the Tous dam was an 36.5-m high earth- and rock-fill dam with a concrete spillway. The site is located on the Eastern coast of Spain. The spillway was controlled by three (3) steel gates (16-m wide, 10.5-m high each). The maximum discharge capacity was 7,000 m^3/s.

On the 20 October 1982, an exceptional flood reached the reservoir. The peak inflow reached 16,000 m^3/s. The electricity supply to the gates had failed as a result of a storm and the spillway gates could not be opened. The dam was overtopped and washed away. Extensive damage resulted from the combination of the extreme flood event, the failure to open the spillway gates and the subsequent dam collapse.

8.3 Safety features for stepped channels

8.3.1 Introduction

A safe design of stepped channels must provide adequate flood discharge facilities, safe channel operation and appropriate control of the water releases. Possible material deterioration must be also taken into account.

A- Adequate spillway facilities

The handling of flood waters can be achieved by the storage of water or water releases (in the main channel or in auxiliary structures).

A safe design procedure must include : 1- an accurate estimation of the Maximum Probable

Flood event, 2- the selection of the maximum channel discharge and eventually 3- the provision for auxiliary flood releases (e.g. fuse spillways, dam overtopping).

B- Safe water release in a stepped channel

The dynamic forces exerted by waters flowing over channels are : 1- the static load of the water, 2- the erosive (or scouring) force resulting from the high velocity pouring over the channel and the bottom shear stress associated with the flow resistance, 3- the abrasion of suspended sediment and 4- the impact force of floating ice, trees and debris of various kinds.

C- Appropriate control of flood release

The channel can be gated or non-gated. Gated spillways can be operated manually or automatically. In any case, the gate operation might be improper : e.g. operator faults caused the Belci dam failure. Another source of failure is the malfunction of the gates : e.g. electrical failure (Tous dam), obstruction by debris (Pinet dam).

Un-gated channels are not dependant upon an opening system. But they might start to operate without warning and induce damage downstream of the structure. They might also be obstructed by debris.

D- Channel material deterioration

From the time of construction and during all the life of the structure, the chute material is subjected to natural hazards. A partial or complete deterioration of the channel material (e.g. concrete, gabion, timber) can be responsible for substantial damage during subsequent channel operations. For example, the failure of the Arizona Canal dam was probably caused by timber deterioration. For concrete channels, the concrete might be deteriorated by chemical reaction (e.g. alkali-aggregate reaction), freezing-thawing (e.g. Lahontan dam), settlement and cracking of the structure, vibration of the structure caused by water surge or equipment operation, erosion of the concrete by cavitation or debris impacts.

8.3.2 Specific problems of stepped channels

The flow patterns of stepped channel flows are very different from smooth chute flow situations and exhibit specific features : e.g., two types of flow regime, cavity recirculation in skimming flow, free-falling jet, eventually followed by hydraulic jump, in nappe flow.

Designers must carefully select the most appropriate flow regime : nappe flow or skimming flow regime. In both cases, the step faces are subjected to mean pressures and pressure fluctuations quite different from smooth channels (see next sub-section). The structural analysis of stepped channels must take into account the additional pressure load.

Further a possible transition from skimming flow to nappe flow (and opposite) should be avoided. Flow conditions near the onset of skimming flow are unstable. The flow instabilities might induce fluctuating hydrodynamic loads and possible vibrations to the hydraulic structure.

Also, for both nappe and skimming flow regimes, an increase of discharge induces a reduction of the rate of energy dissipation. This point must be seriously considered. Indeed, a flood discharge larger than the maximum design could lead to disastrous damage of the downstream stilling basin. The selection process of the Probable Maximum Flood discharge must take into account the risks of failure and the associated costs with the under-estimation of the maximum discharge capacity of the channel.

8.3.3 Hydrodynamic forces on the steps (pressures and pressure fluctuations)

The pressures and pressure fluctuations on the channel steps are important factors affecting the safety of stepped channels. The designer must analyse separately the two cases : i.e., nappe flow and skimming flow regimes.

Nappe flow regime

In a nappe flow situation, large hydrodynamic forces are exerted on the steps (fig. 8-3) : at the impact of the falling nappe, below the hydraulic jump and on the vertical face if the nappe is not adequately ventilated.

On the horizontal step face, the impact of the jet on the step induces large bottom pressures (i.e. larger than hydrostatic) near the impact location. The re-analysis of drop structure experiments suggests that the mean stagnation pressure P_s can be correlated by :

$$\frac{P_s}{\rho_w * g * h} = 1.253 * \left(\frac{d_c}{h}\right)^{0.349} \tag{8-1}$$

Equation (8-1) is compared with experimental data on figure 8-4. Details of the experiments are reported in table 8-2. Presently no information is available on the dynamic pressure fluctuations at drop structures. Experimental results obtained with vertical plunging jets might provide some information on the standard deviation of the dynamic pressures (table 8-3). Further the re-analysis of the results of MAY and WILLOUGHBY (1991) suggests that an order of magnitude of the extreme maximum and minimum pressures at the nappe impact is :

$$P_s + 0.9 * \rho_w * \frac{V_i^2}{2} \qquad \text{instantaneous maximum pressure at nappe impact (8-2)}$$

and the instantaneous minimum pressures are nearly :

$$P_s - 0.6 * \rho_w * \frac{V_i^2}{2} \qquad \text{instantaneous minimum pressure at nappe impact (8-3)}$$

where V_i is the impact velocity of the free-falling nappe (eq. (3-11)).

Downstream of the free-falling nappe, a hydraulic jump takes place on the step. Below hydraulic jumps, the mean pressure is quasi-hydrostatic but large pressure fluctuations are observed (HAGER 1992). The re-analysis of bottom pressure fluctuation records below hydraulic jumps over long periods (table 8-3) indicates that the instantaneous maximum pressures are about :

Fig. 8-3 - Mean pressure on the step in nappe flow regime

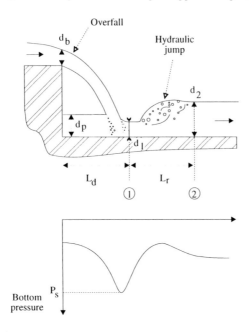

Fig. 8-4 - Mean stagnation pressure at the impact of a free-falling nappe - Experimental data
(ROUSE 1938, MOORE 1943, BAKHMETEFF and FEODOROFF 1943, ROBINSON 1989)

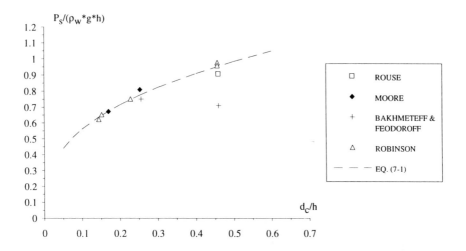

Table 8-2 - Measurements of pressures and pressure fluctuations

Reference	q_w m²/s	d_c/h	d_c/W	Remarks
(1)	(2)	(3)	(4)	(5)
Drop structure ROUSE (1938)		0.459		Aerated nappe. Sharp-crested weir.
MOORE (1943)	0.0673 to 0.123	0.253 to 0.17	0.278 to 0.415	Aerated nappe. W = 0.279 m.
ROBINSON (1989)	0.0623 to 0.124	0.145 to 0.456	0.0803 to 0.127	Aerated nappe. W = 0.914 m. Low tailwater depths.
BAKHMETEFF and FEODOROFF (1943)	0.0556 to 0.134	0.256 to 0.460	0.670 to 1.201	Broad-crested weir with rounded edge. W = 0.102 m.
Vertical plunging jet MAY and WILLOUGHBY (1991)	0.225 to 0.445	N/A	0.864 to 1.36	Bi-dimensional jets (W = 0.2 m). 3.3 < V < 6.6 m/s. Scanning time : 45 minutes.
Hydraulic jump VASILIEV and BUKREYEV (1967)	0.127	N/A		Fr_1 = 5.75. Jump downstream of a sluice gate.
ABDUL KHADER and ELANGO (1973)	0.034 to 0.126	N/A	0.082 to 0.196	4.7 < Fr_1 < 6.6. W = 0.6 m. Jump downstream of a smooth weir. Scanning rate : 100 Hz. Scanning time : 2.5 to 900 s.
TOSO and BOWERS (1988)		N/A		3 < Fr_1 < 10. W = 0.5 m. Jump downstream of inclined slopes (0 to 45 deg.). Scanning rate : 50 Hz. Scanning time : 600 s to 24 hours.
FIOROTTO and RINALDO (1992)	0.0484 to 0.122	N/A	0.207 to 0.384	5 < Fr_1 < 9.5. W = 0.3 m. Scanning rate : 50 to 100 Hz. Scanning time : 164 s to 12 hours.

$$P_{hyd} + 0.6 * \rho_w * \frac{V_1^2}{2} \qquad \text{instantaneous maximum pressure below jump (8-4)}$$

and the instantaneous minimum pressures are nearly :

$$P_{hyd} - 0.4 * \rho_w * \frac{V_1^2}{2} \qquad \text{instantaneous minimum pressure below jump (8-5)}$$

where P_{hyd} is the local hydrostatic pressure and V_1 is the upstream flow velocity. The velocity V_1 can deduced by combining equation (3-6) with the continuity equation. It yields :

$$\frac{V_1}{V_c} = 1.54 * \left(\frac{d_c}{h}\right)^{-0.275} \qquad\qquad (8-6)$$

where V_c is the critical velocity.

It must be noted that, below a hydraulic jump, the extreme minimum pressures (eq. (8-5)) might become negative and could lead to uplift loads under the step bottom.

If the free-falling nappes is not adequately ventilated, subpressures might occur in the nappe cavity and add a suction force on the vertical face of the steps.

In summary, the horizontal face of the steps must be designed to sustain large positive and negative pressures for a wide range of flow conditions. As the location of the falling nappe impact and the position of the hydraulic jump on the step are functions of the discharge (chapter 3), most of the step length must be adequately reinforced to sustain the various loading situations.

Table 8-3 - Pressures and pressure fluctuations at drop structures and below hydraulic jumps

Experimental data	Mean stagnation pressure [a] (Pa)	Std of pressure fluctuations [a] (Pa)	Extreme pressures [a] (Pa)	Remarks
(1)	(2)	(3)	(4)	(5)
Drop structure MOORE (1943), ROBINSON (1989) & ROUSE (1943)	ρ_w*g*h $*1.25*\left(\dfrac{d_c}{h}\right)^{0.35}$			Low tailwater levels.
Vertical plunging jet MAY and WILLOUGHBY (1991)	$P_{hyd} + \dfrac{1}{2}*\rho_w*V_i^2$	$0.1*\dfrac{\rho_w}{2}*V_i^2$	Maximum pressure : $P_{hyd} + 1.9*\dfrac{\rho_w}{2}*V_i^2$ Minimum pressure : $P_{hyd} + 0.4*\dfrac{\rho_w}{2}*V_i^2$	Partially developed jets.
Hydraulic jump VASILIEV and BUKREYEV (1967)		$0.045*\dfrac{\rho_w}{2}*V_1^2$		Partially developed U/S flow.
ABDUL KHADER and ELANGO (1973)	P_{hyd}	$0.085*\dfrac{\rho_w}{2}*V_1^2$	Maximum pressure : $P_{hyd} + 0.26*\dfrac{\rho_w}{2}*V_1^2$ Minimum pressure : $P_{hyd} - 0.2*\dfrac{\rho_w}{2}*V_1^2$	Partially developed U/S flow.
TOSO and BOWERS (1988)	P_{hyd}	$0.07*\dfrac{\rho_w}{2}*V_1^2$	Maximum pressure [b] : $P_{hyd} + 0.6*\dfrac{\rho_w}{2}*V_1^2$ Minimum pressure [b] : $P_{hyd} - 0.4*\dfrac{\rho_w}{2}*V_1^2$	Partially and fully developed U/S flows.
FIOROTTO and RINALDO (1992)	P_{hyd}	$0.08*\dfrac{\rho_w}{2}*V_1^2$ [c]	Maximum pressure [d] : $P_{hyd} + 0.5*\dfrac{\rho_w}{2}*V_1^2$ Minimum pressure [d] : $P_{hyd} - 0.4*\dfrac{\rho_w}{2}*V_1^2$	Partially developed U/S flow.

Notes : [a]   absolute pressure   [b]   scanning time : 24 hours

      [c]   scanning time : 164 s   [d]   scanning time : 12 hours

      Std   standard deviation

      V_i impact velocity

      V_1 upstream flow velocity

      P_{hyd} hydrostatic pressure

Skimming flow regime

In a skimming flow regime, the energy dissipation occurs within the flow by maintaining the eddy recirculation in the step corners beneath the pseudo-bottom formed by the step edges. The mean bottom shear stresses are larger than on smooth spillways. The friction factors range from 0.01 up to 10 (chapter 4). In comparison, f is about 0.01 to 0.05 on smooth concrete channels. Hence, the mean bottom shear stresses (i.e. $\tau_o = f^* \rho_w^* V^2 / 8$) are 10 to 1,000 times larger on stepped chutes than on smooth channels and the steps must be carefully designed to sustain such large shear stresses.

The stresses on the step edges can be reduced by rounding the edges (e.g. Croton Falls dam). Rounded steps induce lower bottom shear stresses but also smaller flow resistance and lower rate energy dissipation.

The efforts and the distributions of pressure and pressure fluctuations on the vertical and horizontal faces of the step are functions of the flow recirculation mechanisms. For each flow pattern (i.e. wake-step interference SK1, wake-wake interference SK2 and stable recirculation SK3 sub-regimes), different hydrodynamic loads are applied to the step faces. The recirculation mechanisms induce fluctuating pressures on both the vertical and horizontal faces of the steps. The dynamic pressure fluctuations may induce local traction and suction.

FRIZELL (1991,1992) and LEJEUNE and LEJEUNE (1994) recorded pressures on the horizontal and vertical faces of steps, using respectively the USBR model (α = 26.6 deg., h = 0.051 m) and the M'Bali model (α = 51.43 deg., h = 0.019 and 0.038 m). In both cases, the results indicated that : (i) the mean pressures were maximum on the horizontal faces of the steps and larger than hydrostatic, (ii) the mean pressures on the vertical faces were less than hydrostatic, and (iii) suction (i.e. pressure below atmospheric) might eventually occur on the vertical faces. Measurements showed also that the mean pressures on the vertical face of the steps decreased with downward sloping steps.

Further experimental work is required to obtain a better understanding of the recirculation process and to predict accurately the pressure distributions. It must be emphasised that only few studies were performed. Additional model and prototype measurements are needed.

8.3.4 Safety of control structures and gates

Channel intakes and gates are very important for a proper flood evacuation and large water releases. They must be always ready to operate and the gate opening must not fail (e.g. Belci and Tous dams). Failure risks of equipment might be reduced by diversifying the energy supply (normal and backup circuits), by providing direct gate operating mechanisms (e.g. engines, crank handle), with periodical maintenance tests and with regular training of the operators.

Obstruction by debris

In any case, the gates and intakes must be accessible and not blocked by ice and debris. The obstruction of spillways by debris was recently highlighted by the failures of the Palagnegra and Pinet dams.

In Tasmania (Australia), where eucalyptus trees grow to over 70-m high, it is usual practice to provide wide spillways with ample overhead clearance (THOMAS 1976). In industrialised areas, debris can range from trees to trucks and camping-cars (e.g. Pinet dam). The channel intakes (and the gates) must be wide and high enough to allow the passage of these debris.

8.4 Discussion

No dramatic accidents were caused directly by the selection of a stepped channel rather than a smooth channel. But GOUBET (1992) addressed some critical safety concerns for the selection of stepped spillways : a spillway structure is designed for 20 to 50 years and might be subjected to exceptional events. E.g. exceptional flood, spillway obstruction, gate failure. How would stepped spillways behave under such extreme conditions ?

Already, it was shown that, for a given spillway geometry, the rate of energy dissipation decreases with increasing discharges. In the event of an exceptional flood leading to an overflow larger than the maximum design discharge, the large residual flow energy might induce substantial damage downstream of the spillway.

Furthermore, in three failure cases (i.e. Arizona canal dam, Minneapolis Mill dam, New Croton dam), failures occur during water spills. And, in each case, the water discharges were much less than the design discharge. What would have been the damage for the design discharge ?

The author wishes to emphasise, again, that the flow conditions near the transition between nappe and skimming flow are unstable and must be avoided. In two failure cases (i.e. Arizona Canal dam, New Croton dam), the spills occurred with flow conditions near the transition nappe/skimming flow. The flow behaviours induced additional fluctuating loads and efforts to the structures and the resulting fatigue might have increase the rate of failure.

In two cases (i.e. New Croton dam, St. Francis dam), large cracks were observed : along the spillway (New Croton dam) or within the dam masonry below the spillway (St. Francis dam). Spillway operation in such conditions could create catastrophic situations. The interactions of the fluctuating hydrodynamic load on the cracks can lead to disastrous circumstances. The New Croton dam and St. Francis dam examples highlight that the design of stepped channels and spillways necessitates a good quality of the construction work.

Further, any hydraulic structure is subjected along the time to natural hazards : e.g. debris impacts, abrasion by sediments, freeze. Stepped spillways are subjected to similar threats (e.g. Lahontan dam, USA).

The author believes that the design of stepped channels and spillways is safe if the designers are perfectly aware of the hydraulic characteristics of stepped chutes. The hydraulics of stepped channels differs substantially from classical smooth chute calculations and it is not taught usually at student levels nor in professional courses. At the present time, only few experienced engineers and researchers have gained the required expertise on stepped channel characteristics. Further research investigations must be carried out to provide new information on the hydrodynamic loads on the steps and the flow behaviour. An increased knowledge of stepped channel flows will increase the safety design of stepped chutes.

8.5 Example of application

Considering a stepped chute (8 steps) with nappe flow regime, the discharge per unit width is 0.3 m^2/s. The dam height is 12 metres. The step height and length are : h = 1.5 m and l = 4 m. Compute the hydraulic characteristics of the flow. Deduce the mean pressures and extreme pressures on the steps.

At each step, the entire flow conditions can be computed using equations (3-1) to (3-28). Assuming that the nappe is adequately ventilated, the pressure on the vertical face of the steps is quasi-atmospheric. On the horizontal face of the steps, the mean pressure the maximum at the impact of the free-falling nappe (eq. (8-1)). The extreme pressures are deduced from plunging jet and hydraulic jump experiments (table 8-3). The main results are summarised in table 8-4.

Discussion

Equation (3-14) indicates that the flow is a nappe flow regime with fully-developed hydraulic jump (regime NA1). Equations (8-1) to (8-5) enable to estimate the stagnation pressure at the impact of the nappe and the extreme pressures at the nappe impact and below the hydraulic jump.

The largest value of the extreme maximum pressures is observed at the nappe impact (i.e. 2.3 m of H_2O at x = L_d). The lowest value of the extreme minimum pressures occurs below the hydraulic jump (i.e. -0.4 m of H_2O). Such a result implies possible suction (i.e. uplift force) on the horizontal face of the steps.

The results are summarised on figure 8-5 where the dimensionless bottom pressures (mean, maximum and minimum extremes) are plotted as a function of the distance along the step.

Table 8-4 -Application for h = 1.5 m, q_w = 0.3 m²/s, H_{dam} = 12 m

Variable (1)	Value (2)	Unit (3)	Eq. (4)	Comments (5)
d_c/h	0.14			
$(d_c/h)_{char}$	0.32		(3-14)	Fully developed hydraulic jump.
d_b	0.15	m	(3-1)	Flow depth at the brink of a step.
d_i	0.056	m	(3-10)	Nappe thickness at the impact with the receiving pool.
V_i	5.4	m/s	(3-11)	Impact velocity of the nappe.
θ	69.4	degrees	(3-12)	Jet angle of the impinging nappe.
d_1	0.066	m	(3-6)	Flow depth upstream of hydraulic jump
V_1	4.56	m/s		Flow velocity upstream of hydraulic jump
Fr_1	5.7			Froude number at start of hydraulic jump.
d_2	0.505	m	(3-7)	Flow depth downstream of hydraulic jump.
d_p	0.409	m	(3-8)	Flow depth in pool beneath the nappe.
L_d	1.31	m	(3-9)	Length of drop.
L_r	2.2	m	(3-13)	Roller length of the hydraulic jump.
Q_{air}^{nappe}/Q_w			(3-17)	Nappe ventilation at each step.
Nappe impact				
P_s	9.26	kPa	(8-1)	Stagnation pressure at nappe impact.
Max. pressure	22.34	kPa	(8-2)	Extreme maximum pressure at nappe impact.
Min. pressure	0.53	kPa	(8-3)	Extreme minimum pressure at nappe impact.
Hydraulic jump				
Max. pressure	11.18	kPa	(8-4)	Extreme maximum pressure below jump.
Min. pressure	-3.51	kPa	(8-5)	Extreme minimum pressure below jump.

Fig. 8-5 - Dimensionless bottom pressures $P/(\rho_w{}^*g{}^*d_c)$ on the horizontal face of steps as a function of the distance along the step x/l for a nappe flow regime
q_w = 0.3 m2/s, h = 1.5 m and l = 4 m (paragraph 8.5)

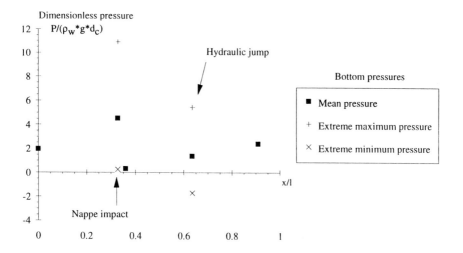

CHAPTER 9
CONCLUSION

Stepped channels have been used since more than 2,500 years. The Assyrian engineers were probably the first to design overflow dams with stepped spillways. Later, Roman, Moslem and Spanish designers used a similar technique. Overflow stepped spillways were selected to contribute to the stability of the dam, for their simplicity of shape and later to reduce flow velocities. Early irrigation systems in Yemen and Peru included also drops and steps to assist in energy dissipation. At the end of the 19-th century, a significant number of dams were built with overflow stepped spillways (SCHUYLER 1909, WEGMANN 1911, KELEN 1933). Most dams were masonry or concrete structures with granite or concrete blocks protecting the downstream face.

Since the beginning of the 20-th century, stepped chutes have been designed more specifically to dissipate flow energy (e.g. New Croton dam). The steps increase significantly the rate of energy dissipation taking place on the channel face and reduce the size of the required downstream energy dissipation and the risks of scouring.

Recently, new construction materials (e.g. RCC, strengthened gabions) have increased the interest for stepped channels and spillways. The construction of stepped chutes is compatible with the slipforming and placing methods of roller compacted concrete and with the construction techniques of gabion dams. Further, some recent advances in the mechanisms of air entrainment and air-water gas transfer enable economical design of aeration cascades for water treatment applications.

The main hydraulic features of stepped channel flows are the two different flow regimes : **nappe flow regime** for small discharges and flat channel slopes, and **skimming flow regime** (table 9-1, fig. 3-2 and 4-11).

In a *nappe flow regime*, the water proceeds in a series of plunges from one step to another. The flow from each step hits the step below as a free-falling jet, eventually followed by a hydraulic jump on the step (sub-regimes NA1 and NA2). Stepped spillways with nappe flows can be analysed as a succession of drop structures. In a nappe flow situation, the head loss at any intermediary step equals the step height. The energy dissipation occurs by jet breakup in air, by jet mixing on the step and with the formation of a hydraulic jump on the step. Most of the flow energy is dissipated on the stepped spillway for large dams. But, for a given dam height, the rate of energy dissipation decreases when the discharge increases.

In nappe flows, air bubble entrainment occurs at the intersection of the falling nappe with the receiving pool and at the toe of the hydraulic jump. The air, entrained by the falling nappe, might induce some sub-atmospheric pressures in the air cavity between the nappe and the vertical step. In such a case, the air cavity must be adequately ventilated to prevent oscillations of the free-falling nappe and the associated noise.

The occurrence of nappe flow regime implies relatively large steps and flat slopes. This situation is not often practical, but it may apply to relatively flat spillways, streams, creeks, river training and storm waterways. For steep channels or small step heights, a skimming flow regime takes place.

In the *skimming flow regime*, the water flows down the stepped face as a coherent stream, skimming over the steps. The external edges of the steps form a pseudo-bottom over which the flow passes. Beneath this, recirculating vortices develop and are maintained through the transmission of shear stress from the waters flowing past the step edges.

For very flat slopes ($\alpha < 27$ degrees), large stable recirculation vortices cannot develop. The recirculating vortices do not fill the entire cavity between the edges, and the wake from one edge interferes with the next step. The flow pattern is characterised by the impact of the wake on the next step, a three-dimensional unstable recirculation in the wake and some friction drag (skin friction) on the step downstream of the wake impact ("wake-step interference" sub-regime SK1). For an increase of slope, the tail of the wake starts interfering with the next wake and the friction drag component disappears ("wake-wake interference" sub-regime SK2). For steep slopes ($\alpha > 27$ degrees), a stable recirculation in the cavities between adjacent steps is observed ("recirculating cavity flow" sub-regime SK3). The recirculating eddies are large two-dimensional and possibly three-dimensional vortices. The energy dissipation and the flow resistance are functions of the energy required to maintain the circulation of these large-scale turbulent vortices.

Skimming flows are characterised by large friction losses. The analysis of model data suggests that the friction factor is independent of the skin roughness, channel slope and Reynolds number. In a wake-step interference regime (SK1), the friction factor increases with increasing relative form roughness k_s/D_H, where k_s is the step depth normal to the flow direction (i.e. $k_s = h^* \cos\alpha$). In a recirculating cavity flow regime (SK3), experimental values of the friction factor range from $f = 0.1$ to 5 with a mean value of about 1.0.

At the upstream end of a stepped chute, the flow is smooth and glassy. However, next to the bottom, turbulence is generated and a boundary layer grows until the outer edge of the boundary layer reaches the free-surface. At that location, called the inception point of air entrainment, the turbulence next to the free surface becomes large enough to overcome surface tension and buoyancy effects, and free-surface aeration occurs. Downstream of the inception point, a layer containing a mixture of both air and water extends gradually through the fluid. Further downstream the flow becomes fully aerated. The air entrainment affects the flow properties, particularly by reducing the flow resistance and the rate of energy dissipation.

Table 9-1 - Summary of flow regimes above stepped chutes

Regime (1)	Description (2)	Flow conditions (3)	Remark (4)
NA1	Nappe flow with fully-developed hydraulic jump	$d_c/h < 0.0916*(h/l)^{-1.276}$	Eq. (3-14)
NA2	Nappe flow with partially-developed hydraulic jump	$d_c/h > 0.0916*(h/l)^{-1.276}$ and $d_c/h < 1.057 - 0.465*h/l$	Eq. (3-14) and (4-1)
NA3	Nappe flow without hydraulic jump	$d_c/h \leq 1.057 - 0.465*h/l$	Steep slope, inclined downward step
SK1	Skimming flow with wake-step interference	$d_c/h > 1.057 - 0.465*h/l$ and $h/l < 0.5$ $$\frac{1}{\sqrt{f}} = 1.42 * Ln\left(\frac{D_H}{k_s}\right) - 1.25$$	Eq. (4-1) Eq. (4-30)
SK2	Skimming flow with wake-wake interference	$d_c/h > 1.057 - 0.465*h/l$ and $h/l \sim 0.5$	
SK3	Skimming flow with recirculating cavity flow	$d_c/h > 1.057 - 0.465*h/l$ and $h/l > 0.5$ $f \sim 1.0$	Eq. (4-1) Fig. 4-10

On long chutes, the rate of energy dissipation is greater with skimming flows than with nappe flows. The maximum energy dissipation is achieved with 30-degree slopes. For steeper channels, the drag reduction induced by the flow aeration reduces the flow resistance and the rate of energy dissipation. For short channels, skimming flows do not reach uniform flow conditions and nappe flows enable larger energy dissipation. It must be emphasised that, with both nappe and skimming flows, an increase of water discharge generates a decrease of the rate of energy dissipation.

In both nappe and skimming flow regimes, the air-water gas transfer taking place above a stepped chute is enhanced by the strong turbulent mixing and by the aeration of the flow. The flow aeration must be taken into account for the re-oxygenation of polluted streams and rivers, but also to explain the high fish mortality downstream of large hydraulic structures.

As the flow patterns of stepped channel flows are very different from smooth chute flow situations, designers must analyse carefully stepped channel flows. Firstly, engineers must select the appropriate flow regime. Then, the hydraulic calculations are developed for that flow situation : nappe flow or skimming flow. Designers must take into account the hydraulic characteristics, the flow aeration and the hydrodynamic interactions between the flow and the steps. In the past, several accidents involved stepped chute structures. On a stepped chute, the step faces are subjected to large mean pressures and pressure fluctuations. And the structural analysis of stepped channels must take into account the additional pressure loads.

Further a possible transition from skimming flow to nappe flow (and opposite) must be avoided. Flow conditions near the onset of skimming flow are unstable. The flow instabilities might induce fluctuating hydrodynamic loads and possible vibrations to the hydraulic structure.

Also, for both nappe and skimming flow regimes, an increase of discharge induces a reduction of the rate of energy dissipation. Such a consideration must be seriously examined in the design stages of a stepped chute. Indeed, a flood release larger than the maximum design discharge could lead to disastrous damage at the downstream end of the structure. For stepped spillways, the selection of the maximum design discharge must take into account the risks of failure and the costs associated with the under-estimation of the maximum discharge capacity.

The author believes that the design of stepped channels and spillways is safe if the designers are perfectly aware of the hydraulic characteristics of stepped chutes. The hydraulics of stepped channels differs substantially from classical smooth chute calculations. It is hoped that, by reading the present monograph, engineers and researchers have gained some expertise on the hydraulic properties of stepped channels.

Future research

Further research investigations must be carried out to provide new information on the flow characteristics of stepped channels. An increased knowledge of stepped channel flows will increase the safety of stepped chutes.

This study has highlighted the lack of expertise and knowledge on :

- the hydraulic characteristics of <u>skimming flows past flat slopes</u>,

- the mechanisms of <u>flow recirculation</u> in skimming flow regime,

- the aerated flow properties of skimming flows, in particular the lack of experimental data taking into account the <u>effects of air entrainment</u>,

- the <u>air-water gas transfer process</u> in nappe flow regime and in skimming flow regime.

For skimming flows, new experimental work is required to provide a better understanding of the flow recirculation in the cavities beneath the pseudo-bottom formed by the step edges. Indeed most of the energy is dissipated in maintaining the recirculation of the cavity.

In a nappe flow regime, additional studies of the air-water gas transfer mechanisms for plunging jets and for hydraulic jumps is necessary. There is presently no information of the distributions of air bubble sizes and of air-water interface area, and the prediction of the downstream dissolved gas content is somehow purely empirical.

REFERENCES

ABDUL KHADER, M.H., and ELANGO, K. (1973). "Turbulent Pressure Field beneath a Hydraulic Jump." *Jl of Hyd. Res.*, IAHR, Vol. 12, No. 4, pp. 469-489.

ABT, S.R., and JOHNSON, T.L. (1991). "Riprap Design for Overtopping Flow." *Jl of Hyd. Engrg.*, ASCE, Vol. 117, No. 8, pp. 959-972.

ADACHI, S. (1964). "On the Artificial Strip Roughness." *Disaster Prevention Research Institute Bulletin*, No. 69, Kyoto University, Japan, March, 20 pages.

AGOSTINI, R., BIZZARRI, A., MASETTI, M., and PAPETTI, A. (1987). "Flexible Gabion and Reno Mattress Structures in River and Stream Training Works. Section One : Weirs." *Officine Maccaferri*, Bologna, Italy, 2nd edition.

AIVAZYAN, O.M. (1986). "Stabilized Aeration on Chutes." *Gidrotekhnicheskoe Stroitel'stvo*, No. 12, pp. 33-40 (in Russian). (Translated in Hydrotechnical Construction, 1987, Plenum Publ., pp. 713-722).

ALLEN, P. (1984). "Modelling Flow Over and Through Overtopped Rockfill Embankments." *M.Eng. thesis*, Dept. of Civil Eng., University of Queensland, Brisbane, Australia.

APHA, AWWA, and WPCF (1985). "Standard Methods for the Examination of Water and Wastewater." *American Public Health Association Publ.*, 6th Ed., 1985.

APHA, AWWA, and WPCF (1989). "Standard Methods for the Examination of Water and Wastewater." *American Public Health Association Publ.*, 7th Ed., 1989.

APTED, R.W., and NOVAK, P. (1973). "Oxygen Uptake at Weirs." *Proc. 15th IAHR Congress*, Vol. 1, Paper No. B23, Istanbul, Turkey, pp. 177-186.

AURAND, C.D. (1986). "Fountains and Pools." *PDA Publ.*, Mesa, USA.

AVDEEV, A.A., DROBKOV, V.P., and KHALME, P.S. (1991). "Turbulent Momentum Transfer in a Bubble Layer." *Teplofizika Vysokikh Temperatur*, Vol. 29, No. 4, pp. 775-780 (in Russian). (Translated in *High Temperature (USSR)*, Vol. 29, 1991, pp. 608-612).

AVERY, S.T., and NOVAK, P. (1975). "Oxygen Uptake in Hydraulic Jumps and at Overfalls." *Proc. 16th IAHR Congress*, Paper C38, Sao Paulo, Brazil, pp. 329-337.

AVERY, S.T., and NOVAK, P. (1978). "Oxygen Transfer at Hydraulic Structures." *Jl of Hyd. Div.*, ASCE, Vol. 104, No. HY11, pp. 1521-1540.

BABB, A.F., and AUS, H.C. (1981). "Measurements of Air in Flowing Water." *Jl of Hyd. Div.*, ASCE, Vol. 107, No. HY12, pp. 1615-1630.

BACK, P.A.A., FREY, J.P., and JOHNSON, G. (1973). "P.K. Le Roux Dam : Spillway Design and Energy Dissipation." *Proc. 11th ICOLD Congress*, Madrid, Spain, Q. 41, R. 76, Vol. II, pp. 1439-1468.

BAKER, R. (1990). "Precast Concrete Blocks for High Velocity Flow Applications." *J. IWEM*, Vol. 4, Dec., pp. 552-557. Discussion : Vol. 4, pp. 557-558.

BAKER, R., and GARDINER, K. (1994). "The Construction and Performance of a Wedge Block

Spillway at Brushes Clough Reservoir." *Proc. British Dam Society Conf.*, Exeter Univ., Sept., UK.

BAKHMETEFF, B.A., and MATZKE, A.E. (1936). "The Hydraulic Jump in Terms of Dynamic Similarity." *Transactions*, ASCE, Vol. 101, pp. 630-647. Discussion : Vol. 101, pp. 648-680.

BARRETT, M.J., GAMESON, L.H., and OGDEN, C.G. (1960). "Aeration Studies at Four Weir Systems." *Water and Water Engrg.*, Sept., pp. 407-413.

BASS, R.P. (1993). "Roller Compacted Concrete Provides Design Alternative for Dam Construction." *Water & Wastewater Intl*, Vol. 8, No. 3, pp. 22-26.

BATHURST, J.C. (1978). "Flow Resistance of Large-Scale Roughness." *Jl of Hyd. Div.*, ASCE, Vol. 104, No. HY12, pp. 1587-1603.

BATHURST, J.C. (1985). "Flow Resistance Estimation in Mountain Rivers." *Jl of Hyd. Engrg.*, ASCE, Vol. 111, No. 4, pp. 625-643.

BATHURST, J.C. (1988). "Velocity Profile in High-Gradient, Boulder-Bed Channels." *Proc. Intl Conf. on Fluvial Hydraulics*, IAHR, May, Budapest, Hungary, pp. 29-34.

BATTISON, E.A. (1975). "Ascutney Gravity-Arch Mill Dam - Windsor, Vermont 1834." *IA Jl of the Soc. of Industrial Archeology*, Summer, Vol. 1, No. 1, pp. 53-58.

BAUER, S.W., and GRAF, W.H. (1971). "Free Overfall as Flow Measuring Device." *Jl of Irrig. and Drain.*, ASCE, Vol. 97, No. IR1, pp. 73-83.

BAYAT, H.O. (1991). "Stepped Spillway Feasibility Investigation." *Proc. 17th ICOLD Congress*, Vienna, Austria, Q. 66, R. 98, pp. 1803-1817.

BAZIN, H. (1865). "Recherches Expérimentales sur l'Ecoulement de l'Eau dans les Canaux Découverts." ('Experimental Research on Water Flow in Open Channels.') *Mémoires présentés par divers savants à l'Académie des Sciences*, Paris, France, Vol. 19, pp. 1-494 (in French).

BEITZ, E., and LAWLESS, M. (1992). "Hydraulic Model Study for dam on GHFL 3791 Isaac River at Burton Gorge." *Water Resources Commission Report*, Ref. No. REP/24.1, Sept., Brisbane, Australia.

BICUDO, J.R., and GIORGETTI, M.F. (1991). "The Effects of Strip Bed Roughness on the Reaeration Rate Coefficient." *Water Sci. Tech.*, Vol. 23, Kyoto, Japan, pp. 1929-1939.

BIN, A.K. (1993). "Gas Entrainment by Plunging Liquid Jets." *Chem. Eng. Science*, Vol. 48, No. 21, pp. 3585-3630.

BINDO, M., GAUTIER, J., and LACROIX, F. (1993). "The Stepped Spillway of M'Bali Dam." *Intl Water Power and Dam Construction*, Vol. 45, No. 1, pp. 35-36.

BINNIE, A.R. (1913). "Rainfall Reservoirs and Water Supply." *Constable & Co*, London, UK, 157 pages.

BINNIE, G.M. (1987). "Early Dam Builders in Britain." *Thomas Telford*, Londond, UK, 181 pages.

BONETTO, F., and LAHEY, R.T. Jr (1993). "An Experimental Study on Air Carryunder due to a Plunging Liquid Jet." *Intl Jl of Multiphase Flow*, Vol. 19, No. 2, pp. 281-294. Discussion : Vol. 20, No. 3, pp. 667-770.

BOUSSINESQ, J.V. (1877). "Essai sur la Théorie des Eaux Courantes." ('Essay on the Theory of

Water Flow.') *Mémoires présentés par divers savants à l'Académie des Sciences*, Paris, France, Vol. 23, ser. 3, No. 1, supplément 24, pp. 1-680 (in French).

BOUYGE, B., GARNIER, G., JENSEN, A., MARTIN, J.P., and STERENBERG, J. (1988). "Construction et Contrôle d'un Barrage en Béton Compacté au Rouleau (BCR) : un Travail d'Equipe." ('Roller Compacted Concrete Dam Construction and Works Supervision : a Team Job.') *Proc. 16th ICOLD Congress*, San Francisco, USA, Q. 62, R. 34, pp. 589-612 (in French).

BOWIE, G.., MILLS, W.B., FORDELLA, D.B., CAMPBELL, C.L., PAGENKOPF, J.R., RUPP, G.L., JOHNSON, K.M., CHAN, P.W.H., CHERINI, S., and CHAMBERLIN, C.E. (1985). "Rates, Constants, and Kinetics Formulations in Surface Water Quality Modeling." *Tetra Tech. Report*, No. EPA/600/3-85-040, 2nd ed., 475 pages.

BOYDEN, B.H., BANH, D.T., and HUCKABAY, H.K. (1990). "Stripping Chlorinated VOCs from Drinking Water via Inclined Cascade Aeration." *Proc. 18th Australasian Chem. Eng. Conf.*, Auckland, NZ, pp. 1024-1031.

BOYER, P.B. (1971). "Gas Supersaturation Problem in the Columbia River." *Proc. Intl Symp. on Man-Made Lakes*, American Geophysical Union, Knoxville TN, USA, pp. 701-705.

BRUSCHIN, J., BAUER, S., DELLEY, P., and TRUCCO, G. (1982). "The Overtopping of the Palagnedra dam." *Intl Water and Dam Construction*, Vol. 34, No. 1, Jan., pp. 13-19.

BUTTS, T.A., and EVANS, R.L. (1983). "Small Stream Channel Dam Aeration Characteristics." *Jl of Envir. Engrg.*, ASCE, Vol. 109, No. 3, pp. 555-573.

CAIN, P. (1978). "Measurements within Self-Aerated Flow on a Large Spillway." *Ph.D. Thesis*, Ref. 78-18, Dept. of Civil Engrg., Univ. of Canterbury, Christchurch, New Zealand.

CAIN, P., and WOOD, I.R. (1981). "Measurements of Self-aerated Flow on a Spillway." *Jl. Hyd. Div.*, ASCE, 107, HY11, pp. 1425-1444.

CASPERSON, L.W. (1993). "Fluttering Fountains." *Jl Sound and Vibration*, Vol. 162, No. 2, pp. 251-262.

CASTELEYN, J.A., KOLKMAN, P.A., and VAN GROEN, P. (1977). "Air Entrainment in Siphons : Results of Tests in Two Scale Models and an Attempt at Extrapolation." *Proc. 17th IAHR Congress*, Baden-Baden, Germany, pp. 499-506 (also : Delft Hydraulic Laboratories, Holland, Report n°187).

CHAMANI, M.R., and RAJARATNAM, N. (1994). "Jet Flow on Stepped Spillways." *Jl of Hyd. Engrg.*, ASCE, Vol. 120, No. 2, pp. 254-259.

CHANSON, H. (1988). "A Study of Air Entrainment and Aeration Devices on a Spillway Model." *Ph.D. thesis*, Ref. 88-8, Dept. of Civil Engrg., University of Canterbury, New Zealand.

CHANSON, H. (1989a). "Study of Air Entrainment and Aeration Devices." *Jl of Hyd. Research*, IAHR, Vol. 27, No. 3, pp. 301-319.

CHANSON, H. (1989b). "Flow downstream of an Aerator. Aerator Spacing." *Jl of Hyd. Research*, IAHR, Vol. 27, No. 4, pp. 519-536.

CHANSON, H. (1992a). "Air Entrainment in Chutes and Spillways." *Research Report No. CE 133*, Dept. of Civil Engineering, University of Queensland, Australia, Feb., 85 pages.

CHANSON, H. (1992b) "Drag Reduction in Self-Aerated Flows. Analogy with Dilute Polymer Solutions and Sediment Laden Flows" *Research Report No. CE141*, Dept. of Civil Engineering, University of Queensland, Australia, Oct., 28 pages.

CHANSON, H. (1992c). "Reduction of Cavitation on Spillways by induced Air Entrainment - Discussion." *Can. Jl of Civ. Eng.*, Vol. 20, Oct., pp. 926-928.

CHANSON, H. (1993a). "Self-Aerated Flows on Chutes and Spillways." *Jl of Hyd. Engrg.*, ASCE, Vol. 119, No. 2, pp. 220-243. Discussion : Vol. 120, No. 6, pp. 778-782.

CHANSON, H. (1993b). "Velocity Measurements within High Velocity Air-Water Jets." *Jl of Hyd. Res.*, IAHR, Vol. 31, No. 3.

CHANSON, H. (1993c). "Stepped Spillway Flows and Air Entrainment." *Can. Jl of Civil Eng.*, Vol. 20, No. 3, June, pp. 422-435.

CHANSON, H. (1993d). "Environmental Impact of Large Water Releases in Chutes : Oxygen and Nitrogen Contents due to Self-Aeration." *Proc. 25th IAHR Congress*, Tokyo, Japan, paper D9-1, Vol. 5, pp. 273-280.

CHANSON, H. (1993e). "Characteristics of Undular Hydraulic Jumps." *Research Report No. CE146*, Dept. of Civil Engineering, University of Queensland, Australia Nov..

CHANSON, H., and CUMMINGS, P.D. (1992). "Aeration of the Ocean due to Plunging Breaking Waves." *Research Report No. CE142*, Dept. of Civil Engineering, University of Queensland, Australia, Nov., 42 pages.

CHANSON, H., and QIAO, G.L. (1994). "Air Bubble Entrainment and Gas Transfer at Hydraulic Jumps." *Research Report No. CE149*, Dept. of Civil Engineering, University of Queensland, Australia, Aug., 68 pages.

CHEN, C.L. (1990). "Unified Theory on Power Laws for Flow Resistance." *Jl of Hyd. Engrg.*, ASCE, Vol. 117, No. 3, pp. 371-389.

CHRISTODOULOU, G. C. (1993). "Energy Dissipation on Stepped Spillways." *Jl of Hyd. Engrg.*, ASCE, Vol. 119, No. 5, pp. 644-650.

CIRPKA, O., REICHERT, P., WANNER, O., MÜLLER, S.R., and SCHWARZENBACH, R.P. (1993). "Gas Exchange at River Cascades : Field Experiments and Model Calculations." *Environmental Science Technology*, Vol. 27, No. 10, pp. 2086-2097.

CLAY, P.H. (1940). "The Mechanism of Emulsion Formation in Turbulent Flow." *Proc. Ro. Acad. Sci. (Amsterdam)*, Vol. 43, Part I : pp. 852-865 & Part II : pp. 979-990.

COLLETT, K.O. (1975). "Unusual Surfaces for Large Spillways." *ANCOLD Bulletin*, No. 42, July, pp. 3-10.

COMOLET, R. (1979a). "Sur le Mouvement d'une bulle de gaz dans un liquide." ('Gas bubble motion in a liquid medium.') *Jl La Houille Blanche*, 1979, No. 1, pp. 31-42 (in French).

COMOLET, R. (1979b). "Vitesse d'ascension d'une bulle de gaz isolée dans un liquide peu visqueux." ('The terminal Velocity of a Gas Bubble in a Liquid of very low Viscosity.') *Jl de Mécanique Appliquée*, Vol. 3, No. 2, pp. 145-171 (in French).

Concrete (1993). "Clywedog Dam - Mid-Wales." *Concrete*, Vol. 27, No. 5, pp. 27-29.

COOK, O.F. (1916). "Staircase Farms of the Ancients." *National Geographic Magazine*, Vol. 29, pp. 474-534

CORIOLIS, G.G. (1836). "Sur l'établissement de la formule qui donne la figure des remous et sur la correction qu'on doit introduire pour tenir compte des différences de vitesses dans les divers points d'une même section d'un courant." ('On the establishment of the formula giving the backwater curves and on the correction to be introduced to take into account the velocity differences at various points in a cross-section of a stream.') *Annales des Ponts et Chaussées*, 1st Semester, Series 1, Vol. 11 (in French).

CORSI, R.L., SHEPHERD, J., KALICH, L., MONTEITH, H., and MELCER, H. (1992). "Oxygen Transfer and VOC Emissions from Sewer Drop Structures." *Proc. 1992 Nat. Conf. on Hydraulic Engrg.*, Water Forum '92, ASCE, Baltimore, USA, pp. 305-310.

CRAYA, A. (1948). "Hauteur d'eau à l'Extrémité d'un Long Déversoir." ('Flow Depth at the Downstream End of a Long Spillway.') *Jl La Houille Blanche*, Mar.-Apr., pp. 185-186 (in French).

CREAGER, W.P. (1917). "Engineering of Masonry Dams." *John Wiley & Sons*, New York, USA.

CREAGER, W.P., JUSTIN, J.D., and HINDS, J. (1945). "Engineering for Dams." *John Wiley & Sons*, New York, USA, 3 Volumes.

CROCKER, N. (1987). "Fountains." *Landscape Australia*, No. 2, pp. 137-154, No 3, pp. 201-206, No. 4, pp. 324-332.

CROUCH, D.P. (1991). "Spanish Water Technology in New Spain - Transfer and Alterations." *8th Intl Symp. on History of Hydraulics*, Merida, Spain (Mitteilungen, Inst. für Wasserbau der Technische Univ. Braunschweig, Germany, Col. 117, 1992, pp. 180-227).

CUMMINGS, P.D. (1994). "Gas Transfer at Overfalls and Drop Structures." *Proc. Intl Conf. on Hydraulics in Civil Eng.*, IEAust., Brisbane, Australia, 15-17 Feb., pp. 105-109.

CUMMINS, P.J., SMITH, B.R., and EVANS, R.C. (1985). "Rehabilitation of Goulburn Weir Foundations and Superstructures." *Proc. 15th ICOLD Congress*, Lausanne, Switzerland, Q. 59, R. 38, pp. 603-618.

CURTIS, R.P., and LAWSON, J.D. (1967). "Flow over and through Rockfill Banks." *Jl of Hyd. Div.*, ASCE, Vol. 93, No. HY5, pp. 1-21.

DANIIL, E.I., and GULLIVER J. (1988). "Temperature Dependence of Liquid Film Coefficient for Gas Transfer." *Jl of Envir. Engrg.*, ASCE, Vol. 114, No. 5, pp. 1224-1229.

DANIIL, E.I., GULLIVER J., and THENE, J.R. (1991). "Water-Quality Assessment for Hydropower." *Jl of Envir. Engrg.*, ASCE, Vol. 117, No. 2, pp. 179-193.

DEGOUTTE, G., PEYRAS, L., and ROYET, P. (1992). "Skimming Flow in Stepped Spillways - Discussion." *Jl of Hyd. Engrg.*, ASCE, Vol. 118, No. 1, pp. 111-114.

DEGREMONT (1979). "Water Treatment Handbook" *Halsted Press Book*, John Wiley & Sons, 5th edition, New York, USA.

DELAGRAVE, A., MARCHAND, J., PIGEON, M., RANC, R., and MARZIN, J. (1994). "Résistance au Gel-Dégel du Béton Compacté au Rouleau pour les Barrages à Base de liant

Rolac." ('Freeze-thaw Durability of Roller-compacted Concrete made with Rolac Binder.') *Materials and Structures*, Vol. 27, No. 165, pp. 26-32 (in French).

Department of the Environment (1973). "Aeration at Weirs." *Notes on Water Pollution*, No. 61, Water Res. Lab., Elder Way, Stevenage, Herts, England.

DETSCH, R.M., and SHARMA, R.N. (1990). "The Critical Angle for Gas Bubble Entrainment by Plunging Liquid Jets." *Chem. Eng. Jl*, Vol. 44, pp. 157-166.

DETSCH, R.M., and STONE, T.A. (1992). "Air-Entrainment and Bubble Spectra of Plunging Liquid Jets at Acute Angles." *AIChE Symposium Series*, No. 286, Vol. 88, pp. 119-125.

DIACON, A., STRMATIU, D., and MIRCEA, N. (1992). "An Analysis of the Belci Dam Failure." *Intl Water Power & Dam Construction*, Vol. 44, No. 9, Sept., pp. 67-72.

DIEZ-CASCON, J., BLANCO, J.L., REVILLA, J., and GARCIA, R. (1991). "Studies on the Hydraulic Behaviour of Stepped Spillways." *Intl Water Power and Dam Construction*, Vol. 43, No. 9, Sept., pp. 22-26.

DOLEN, T.P., and von FAY, K.F. (1993). "The Development of Roller Compacted Concrete Mixtures for Bureau of Reclamation Mass Concrete Construction." *American Concrete Institute*, SP-141, USA, LIU and HOFF ed., pp. 133-163.

DUSSART, J., ROBINO, M, BESSIERE, C., and GUERINET, M. (1993). "Le Béton Compacté au Rouleau (BCR) du Barrage de Petit-Saut (Guyane)." ('The Roller Compacted Concrete of the Petit-Saut Dam (Guyana).') *Travaux*, No. 688, June, pp. 14-21 (in French).

EHRENBERGER, R. (1926). "Wasserbewegung in steilen Rinnen (Susstennen) mit besonderer Berucksichtigung der Selbstbelüftung." ('Flow of Water in Steep Chutes with Special Reference to Self-aeration.') *Zeitschrift des Österreichischer Ingenieur und Architektverein*, No. 15/16 and 17/18 (in German) (translated by Wilsey, E.F., U.S. Bureau of Reclamation).

ELLIS, J. (1989). "Guide to Analysis of Open-Channel Spillway Flow." *CIRIA Technical Note No. 134*, 2nd edition, London, UK.

ELMORE, H.L., and WEST, W.F. (1961). "Effect of Water Temperature on Stream Reaeration." *Jl of Sanitary Engrg.*, ASCE, Vol. 87, No. SA6, pp. 59-71.

The Engineer (1939). "The Ladybower Reservoir." *The Engineer*, Vol. 168, pp. 440-442.

Engineering (1966). "Britain's Highest Dam." *Engineering*, Vol. 202, pp. 755-759.

Engineering News. (1905a). "Construction, Repairs and Subsequent Partial Destruction of the Arizona Canal Dam, near Phoenix, Arizona." *Engineering News*, Vol. 53, No. 17, 27 April, pp. 450-451.

Engineering News. (1905b). "Debris Barrier No. 1, Yuba River, California." *Engineering News*, Vol. 53, No. 24, 15 June, pp. 609-610.

ERBE, J. (1974). "Handbook for Designing Decorative Fountains." *Roman Fountains Inc.*, Albuquerque NM, USA.

ERVINE, D.A., and AHMED, A.A. (1982). "A Scaling Relationship for a Two-Dimensional Vertical Dropshaft." *Proc. Intl. Conf. on Hydraulic Modelling of Civil Engineering Structures*, BHRA Fluid Engrg., Coventry, UK, paper E1, pp. 195-214.

ERVINE, D.A., and ELSAWY, E.M. (1975). "The Effect of Falling Nappe on River Aeration." *Proc. 16th IAHR Congress*, Sao Paulo, Brazil, Vol. 3, paper C45, pp. 390-397.

ERVINE, D.A., and FALVEY, H.T. (1987). "Behaviour of Turbulent Water Jets in the Atmosphere and in Plunge Pools." *Proc. Instn Civ. Engrs.*, Part 2, Mar. 1987, 83, pp. 295-314. Discussion : Part 2, Mar.-June 1988, 85, pp. 359-363.

ERVINE, D.A., McKEOGH, E.J., and, ELSAWY, E.M. (1980). "Effect of Turbulence Intensity on the rate of Air Entrainment by Plunging Water Jets." *Proc. Instn Civ. Engrs*, Part 2, June, pp. 425-445.

ESSERY, I.T.S., and HORNER, M.W. (1978). "The Hydraulic Design of Stepped Spillways." *CIRIA Report No. 33*, 2nd edition, Jan., London, UK.

ESSERY, I.T.S., TEBBUTT, T.H.Y., and RASARATNAM, S.K. (1978). "Design of Spillways for Re-aeration of Polluted Waters." *CIRIA Report No. 72*, Jan., London, UK, 36 pages.

ETCHEVERRY, B.A. (1916). "Irrigation Practice and Engineering." *McGraw Hill Book*, New York, USA, 3 Volumes.

EVANS, R.C. (1984). "Goulburn Weir Remodelling." *ANCOLD Bulletin*, No. 70, Dec., pp. 36-38.

EVANS, G.M., JAMESON, G.J., and ATKINSON, B.W. (1992). "Prediction of the Bubble Size Generated by a Plunging Liquid Jet Bubble Column." *Chem. Eng. Sc.*, Vol. 47, No. 13/14, pp. 3265-3272.

FALVEY, H.T. (1980). "Air-Water Flow in Hydraulic Structures." *USBR Engrg. Monograph*, No. 41, Denver, Colorado, USA.

FARRINGTON, I.S. (1980). "The Archaeology of Irrigation Canals with Special Reference to Peru." *World Archaeology*, Vol. 11, No. 3, pp. 287-305.

FARRINGTON, I.S., and PARK, C.C. (1978). "Hydraulic Engineering and Irrigation Agriculture in the Moche Valley, Peru : c. A.D. 1250-1532." *Jl of Archaeological Science*, Vol. 5, pp. 255-268.

FERRELL, R.T., and HIMMELBLAU, D.M. (1967). "Diffusion Coefficients of Nitrogen and Oxygen in Water." *Jl of Chem. Eng. Data*, Vol. 12, No. 1, pp. 111-115.

FERRO, V. (1992). "Flow Measurement with Rectangular Free Overfall." *Jl of Irrig. and Drain.*, ASCE, Vol. 118, No. 6, pp. 956-964.

FERRO, V., and BAIAMONTE, G. (1994). "Flow Velocity Profiles in Gravel-Bed Rivers." *Jl of Hyd. Engrg.*, ASCE, Vol. 120, No. 1, pp. 60-80.

FIOROTTO, V., and RINALDO, A. (1992). "Turbulent Pressure Fluctuations under Hydraulic Jumps." *Jl of Hyd. Res.*, IAHR, Vol. 30, No. 4, pp. 499-520.

FORBES, R.J. (1955). "Studies in Ancient Technology." *Leiden*, E.J. Brill, 9 Vol..

FORD, A.C. (1957). "Repair of Spillway leaks at New Croton, N.Y.." *Jl American Waterworks Association*, Feb., pp. 198-204.

FOREE, E.G. (1976). "Reaeration and Velocity Prediction for Small Streams." *Jl of Envir. Engrg.*, ASCE, Vol. 102, No. EE5, pp. 937-952.

FRIZELL, K.H. (1992). "Hydraulics of Stepped Spillways for RCC Dams and Dam Rehabilitations. " *Proc. 3rd Specialty Conf. on Roller Compacted Concrete*, ASCE, San Diego CA,

USA, pp. 423-439.

FRIZELL, K.H., and MEFFORD, B.W. (1991). "Designing Spillways to Prevent Cavitation Damage." *Concrete International*, Vol. 13, No. 5, pp. 58-64.

FURUYA, Y., MIYATA, M., nd FUJITA, H. (1976). "Turbulent Boundary Layer and Flow Resistance on Plates Roughened by Wires." *Jl of Fluids Eng.*, Trans. ASME, Vol. 98, pp. 635-644.

GAMESON, A.L.H. (1957). "Weirs and the Aeration of Rivers." *Jl Inst. Water Eng. Sc.*, Vol. 11, pp. 477-490.

GAMESON, A.L.H., VANDYKE, K.G., and OGDEN, C.G. (1958). "The Effect of Temperature on Aeration at Weirs." *Water and Water Engrg.*, Nov., pp. 489-492.

GARCIA-DIEGO, J.A. (1977). "Old Dams in Extramadura." *History of Technology*, Vol. 2, pp. 95-124.

GASPAROTTO, R. (1991). "Urban Waterfalls Combine Aesthetics with Water Quality Improvement." *Water Quality Intl.*, No. 4, pp. 36-39.

GASPAROTTO, R. (1992). "Waterfall Aeration Works." *Civil Engineering*, ASCE, Oct., pp. 52-54.

GAUSMANN, R.W., and MADDEN, C.M. (1923). "Experiments with Models of the Gilboa Dam and Spillway." *Transactions*, ASCE, Vol. 86, pp. 280-305.

GEVORKYAN, S.G., and KALANTAROVA, Z.K. (1992). "Design of Tsunami-Protective Embankments with a Complex Front Surface." *Gidrotekhnicheskoe Stroitel'stvo*, No. 5, May, pp. 19-21 (Hydrotechnical Construction, 1992, Plenum Publ., pp. 286-290).

GILL, M.A. (1979). "Hydraulics of Rectangular Vertical Drop Structures." *Jl of Hyd. Res.*, IAHR, Vol. 17, No. 4, pp. 289-302.

GOERTLER, H. (1942). "Berechnung von Aufgaben der freien Turbulenz auf Grund eines neuen Näherungsansatzes." *Z.A.M.M.*, 22, pp. 244-254 (in German).

GOLDRING, B.T., MAWER, W.T., and THOMAS, N. (1980). "Level Surges in the Circulating Water Downshaft of Large Generating Stations." *Proc. 3rd Intl Conf. on Pressure Surges*, BHRA Fluid Eng., F2, Canterbury, UK, pp. 279-300.

GOMEZ, B. (1993). "Roughness of Stable, Armored Gravel Beds." *Water Resources Res.*, Vol. 29, No. 11, pp. 36-31-3642.

GOMEZ-PEREZ, F. (1942). "Mexican Irrigation in the Sixteenth Century." *Civil Engineering*, ASCE, Vol. 12, No. 1, pp. 24-27.

GOUBET, A. (1992). "Evacuateurs de Crues en marches d'Escalier." ('Stepped Spillways.') *Jl La Houille Blanche*, No. 2/3, pp. 159-162. Discussion : No. 2/3, pp. 247-248 (in French).

GRINCHUK, A.S., PRAVDIVETS, Y.P., and SHEKHTMAN, N.V. (1977). "Test of Earth Slope Revetments Permitting Flow of Water at Large Specific Discharges." *Gidrotekhnicheskoe Stroitel'stvo*, No. 4, pp. 22-26 (in Russian). (Translated in Hydrotechnical Construction, 1978, Plenum Publ., pp. 367-373).

GULLIVER, J.S., and HALVERSON, M.J. (1989). "Air-Water Gas Transfer in Open Channels." *Water Ressources Res.*, Vol. 25, No. 8, pp. 1783-1793.

GULLIVER, J.S., THENE, J.R., and RINDELS, A.J. (1990). "Indexing Gas Transfer in Self-Aerated Flows." *Jl of Environm. Engrg.*, ASCE, Vol. 116, No. 3, pp. 503-523. Discussion, Vol. 117, pp. 866-869.

HAGER, W.H. (1983). "Hydraulics of Plane Free Overfall." *Jl of Hyd. Engrg.*, ASCE, Vl. 109, No. 12, pp. 1683-1697.

HAGER, W.H. (1992). "Energy Dissipators and Hydraulic Jump. " *Kluwer Academic Publ.*, Water Science and Technology Library, Vol. 8, Dordrecht, The Netherlands , 288 pages.

HALBRONN, G., DURAND, R., and COHEN DE LARA, G. (1953). "Air Entrainment in Steeply Sloping Flumes." *Proc. 5th IAHR Congress*, IAHR-ASCE, Minneapolis, USA, pp. 455-466.

HALL, L.S. (1943). "Open Channel Flow at High Velocities." *Trans. ASCE*, Vol. 108, pp. 1394-1434.

HALL, W.D. (1993). "Dams and Hydropower in Tennessee Valley." *Intl Water Power and Dam Construction*, Vol. 45, No. 8, Aug., pp. 47-49.

HARTUNG, F., and SCHEUERLEIN, H. (1970). "Design of Overflow Rockfill Dams." *Proc. 10th ICOLD Congress*, Montréal, Canada, Q. 36, R. 35, pp. 587-598.

HAUGEN, H.L., and DHANAK, A.M. (1966). "Momentum Transfer in Turbulent Separated Flow past a Rectangular Cavity." *Jl of Applied Mech.*, Trans. ASME, Sept., pp. 641-464.

HAUSER, G.E., SHANE, R.M., NIZNIK, J.A., and BROCK, W.G. (1992). "Innovative Reregulation Weirs." *Civil Engineering*, ASCE, Vol. 62, No. 5, pp. 64-66.

HAYDUK, W., and LAUDIE, H. (1974). "Prediction of Diffusion Coefficients for Nonelectrolytes in Dilute Aqueous Solutions." *AIChE Jl*, Vol. 20, No. 3, pp. 611-615.

HENDERSON, F.M. (1966). "Open Channel Flow." *MacMillan Company*, New York, USA.

HEWLETT, H.W.M., BOORMAN, L.A., and BRAMLEY, M.E. (1987). "Design of Reinforced Grass Waterways." *CIRIA Report No. 116*, London, UK, 116 pages.

HIGBIE, R. (1935). "Rate of Absorption of a Gas into a Still Liquid." *Trans. AIChE*, Vol. 31, pp. 365-389.

HINDS, J. (1932). "200-Year-Old Masonry Dams in Use in Mexico." *Engineering News-Record*, Vol. 109, Sept. 1, pp. 251-253.

HINDS, J. (1953). "Continuous Development of Dams since 1850." *Transactions*, ASCE, Vol. CT, pp. 489-520.

HINZE, J.O. (1955). "Fundamentals of the Hydrodynamic Mechanism of Splitting in Dispersion Processes." *Jl of AIChE*, Vol. 1, No. 3, pp. 289-295.

HODGE, F.W. (1893). "Prehistoric Irrigation in Arizona." *The American Anthropologist*, Vol. VI, pp. 323-330.

HOLLER, A.G. (1971). "The Mechanism Describing Oxygen Transfer from the Atmosphere to Discharge through Hydraulic Structures. " *Proc. 14th IAHR Congress*, Vol. 1, Paper A45, Paris, France, pp. 372-382.

HOLLINGWORTH, F., and DRUYTS, F.H.W. (1986). "Rollcrete : Some Applications to Dams in South Africa." *Intl Water Power and Dam Construction*, Vol. 38, No. 1, Jan., pp. 13-16.

HORNER, M.W. (1969). "An Analysis of Flow on Cascades of Steps." *Ph.D. thesis*, Univ. of

Birmingham, UK.

HOUSTON, K.L., and RICHARDSON, A.T. (1988). "Energy Dissipation Characteristics of a Stepped Spillway for a RCC Dam." *Proc. Intl Symp. on Hydraulics for High Dams*, IAHR, Beijing, China, pp. 91-98.

HUANXIONG, Xia, and CAIYAN, Han (1991). "Stability of Protection Gabions on the Downstream Slope of Overflow Rockfill Cofferdams." *Proc. 17th ICOLD Congress*, Vienna, Austria, Q. 67, R. 31, pp. 507-523.

HUNGR, O., MORGAN, G.C., VANDINE, D.F., and LISTER, D.R. (1987). "Debris Flow Defenses in British Columbia." *Reviews in Engineering Geology*, Geological Soc. of America, Vol. VII, pp. 20-222.

INNEREBNER, K. (1924). "Overflow Channels from Surge Tanks." *World Power Conference*, 1st, Vol. 2, pp. 481-486.

International Organization for Standardization (1979). "Units of Measurements." *ISO Standards Handbook*, No. 2, Switzerland.

JAMME, G. (1974). "Travaux Fluviaux." ('Fluvial Works.') *Eyrolles*, EDF-DRE collection, Paris, France, 163 pages.

JANSEN, R.B. (1983). "Dams and Public Safety." *US Govt. Printing Office*, US Bureau of Reclamation, Denver, USA, 332 pages.

JARVIS, P.J. (1970). "A Study of the Mechanics of Aeration at Weirs." *Masters thesis*, University of Newcastle upon Tyne, UK.

JELLICOE, S., and JELLICOE, G. (1971). "Water. The Use of Water in Landscape Architecture." *Adam&Charles Black*, London, UK.

JEVDJEVICH, V., and LEVIN, L. (1953). "Entrainment of Air in flowing Water and Technical Problems connected with it." *Proc. 5th IAHR Congress*, IAHR-ASCE, Minneapolis, USA, pp. 439-454.

JOHNSON, P.L. (1984). "Prediction of Dissolved Gas Transfer in Spillway and Outlet Works Stilling Basin Flows." *Proc. 1st Intl Symp on Gas Transfer at Water Surfaces*, Gas Transfer at Water Surfaces, W. BRUTSAERT and G.H. JIRKA ed., pp. 605-612.

JOHNSON, N.A., and TEMPLAR, J. (1974). "Lowering the Warren Dam Spillway." *ANCOLD Bulletin*, No. 44, May, pp. 14-19.

JUDD, H.E., and PETERSON, D.F. (1969). "Hydraulics of Large Bed Element Channels." *UWRL Research Report*, PRWG 17-6, Utah State University, Logan, USA, 115 pages.

JUN, K.S., and JAIN, S.C. (1993). "Oxygen Transfer in Bubbly Turbulent Shear Flow." *Jl of Hyd. Engrg.*, ASCE, Vol. 119, No. 1, pp. 21-36. Discussion : Vol. 120, No. 6, pp. 774-777.

KALINSKE, A.A., and ROBERTSON, J.M. (1943). "Closed Conduit Flow." *Transactions*, ASCE, Vol. 108, pp. 1435-1447.

KAWASE, Y., and MOO-YOUNG, M. (1992). "Correlations for Liquid-Phase Mass Transfer Coefficients in Bubble Column Reactors with Newtonian and Non-Newtonian Fluids." *Can. Jl of Chem. Eng.*, Vol. 70, Feb., pp. 48-54.

KELEN, N. (1933). 'Gewichtsstaumauern und Massive Wehre." ('Gravity Dams and Large Weirs.') *Verlag von Julius Springer*, Berlin, Germany (in German).

KELLER, R.J., and RASTOGI, A.K. (1977). "Design Chart for Predicting Critical Point on Spillways." *Jl of Hyd. Div.*, ASCE, Vol. 103, No. HY12, pp. 1417-1429.

KELLS, J.A. (1993). "Spatially Varied Flow over Rockfill Embankments." *Can. Jl of Civ. Engrg.*, Vol. 20, pp. 820-827. Discussion : Vol. 21, No. 1, pp. 161-166.

KILLEN, J.M. (1982). "Maximum Stable Bubble Size and Associated Noise Spectra in a Turbulent Boundary Layer." *Proc. Cavitation and Polyphase Flow Forum*, ASME, pp. 1-3.

KISTLER, A.L., and TAN, F.C. (1967). "Some Properties of Turbulent Separated Flows." *Physics of Fluids*, Vol. 10, No. 9, Pt II, pp. S165-S173.

KNAUSS, J. (1979). "Computation of Maximum Discharge at Overflow Rockfill Dams (a comparison of different model test results)." *Proc. 13th ICOLD Congress*, New Delhi, India, Q. 50, R. 9, pp. 143-159.

KNIGHT, R.G. (1938). "The Subsidence of a Rockfill dam and the Remedial Measures employed at Eildon Reservoir, Australia." *Proc. Instn Civ. Engrs.*, UK, Vol. 8, 1937/38, pp. 111-191. Discussion : Vol. 8, pp. 192-208 & Vol. 9, pp. 451-495.

KNIGHT, D.W., and MACDONALD, J.A. (1979). "Hydraulic Resistance of Artificial Strip Roughness." *Jl of Hyd. Div.*, ASCE, Vol. 105, No. HY6, June, pp. 675-690.

KOGA, M. (1982). "Bubble Entrainment in Breaking Wind Waves." *Tellus*, Vol. 34, No. 5, pp. 481-489

KRAIJENHOFF, D.A., and DOMMERHOLT, A. (1977). "Brink Depth Method in Rectangular Channel." *Jl of Irrig. and Drain.*, ASCE, Vol. 103, No. IR2, pp. 171-177.

KREST'YANINOV, A.M., and PRAVDIVETS, Y.P. (1986). "Stepped Spillways for Small Dams." *Gidrotekhnicheskoe Melioratsiya*, No. 8, pp. 27-30 (in Russian).

KUMAR, S., NIKITOPOULOS, D.N., and MICHAELIDES, E.E. (1989). "Effect of Bubbles on the Turbulence near the Exit of a Liquid jet." *Experiments in Fluids*, Vol. 7, pp. 487-494.

KUSABIRAKI, D., MUROTA, M., OHNO, S., YAMAGIWA, K., YASUDA, M., and OHKAWA, A. (1990). "Gas Entrainment Rate and Flow Pattern in a Plunging Liquid Jet Aeration System using Inclined Nozzles." *Jl. of Chem. Eng. of Japan*, Vol. 23, No. 6, pp. 704-710.

LARA, P. (1979). "Onset of Air Entrainment for a Water Jet Impinging Vertically on a Water Surface." *Chem. Eng. Sc.*, Vol. 34, pp. 1164-1165.

LAWSON, J.D. (1987). "Protection of Rockfill Dams and Cofferdams against Overflow and Throughflow. The Australian Experience." *Civil Engrg Trans. I.E.Aust.*, Vol. CE29, No. 3, pp. 138-147.

LeBARON BOWEN Jr, R. and ALBRIGHT, F.P. (1958). "Archaelogical Discoveries in South Arabia." *The John Hopkins Press*, Baltimore, USA.

LEFRANC, M. (1992). "Evolution dans l'Exploitation des Evacuateurs de Crues et Rénovations Récentes sur les Barrages d'Electricité de France." ('Evolution in Spillway Operations and Recent Refurbishings on the Dams of the French Electricity Commission.') *Jl La Houille*

Blanche, No 2/3, pp. 163-174. Discussion, p. 248.

LEJEUNE, A., and LEJEUNE, M. (1994). "Some Considerations on the Hydraulic Behaviour of Stepped Spillways." *Proc. Intl Conf. Modelling, Testing and Monitoring for Hydo Powerplants*, UNESCO-IAHR, Budapest, Hungary, July.

LEMPERIERE, F. (1991). "Overspill Rockfill Dams : Conventional and Unconventional Designs. Technology - Costs - Safety." *Proc. 17th ICOLD Congress*, Vienna, Austria, Q. 67, R. 7, pp. 111-127.

LEMPERIERE, F. (1993). "Dams that have Failed by Flooding : an Analysis of 70 Failures." *Intl Water Power and Dam Construction*, Vol. 45, No. 9/10, pp. 19-24.

LEUTHEUSSER, H.J., RESCH, F.J., and ALEMU, S. (1973). "Water Quality Enhancement Through Hydraulic Aeration." *Proc. 15th IAHR Congress*, Istanbul, Turkey, Vol. 2, paper B22, p. 167-175.

LEVIN, L. (1955). "Quelques Réflexions sur la Mécanique de l'écoulement des Mélanges d'Eau et d'Air." ('Notes on the Flow Mechanics of Water-Air Mixtures.') *Jl La Houille Blanche*, No. 4, Aug.-Sept., 1955, pp. 55-557 (in French).

LEVIN, L. (1968). "Formulaire des Conduites Forcées, Oléoducs et Conduits d'Aération." ('Handbook of Pipes, Pipelines and Ventilation Shafts.') *Dunod*, Paris, France (in French).

LEWIS, D.A., and DAVIDSON, J.F. (1982). "Bubble Splitting in Shear Flow." *Trans. IChemE*, Vol. 60, pp. 283-291.

LIEPMANN, H.W., and LAUFER, J. (1947). "Investigation of Free Turbulent Mixing." *NACA Tech. Note*, No. 1257, Aug..

LIN, T.J., and DONNELLY, H.G. (1966). "Gas Bubble Entrainment by Plunging Laminar Liquid Jets." *AIChE Jl*, Vol. 12, No. 3, pp. 563-571.

LYSNE, D.K. (1992). "New Concepts in Spillway Design." *Proc. 2nd Intl Conf. on Hydropower*, LilleHammer, Norway, June, Ed. E. BROCH and D.K. LYSNE, Balkema Publ., pp. 415-423.

McKEOGH, E.J. (1978). "A Study of Air Entrainment using Plunging Water Jets." *Ph.D. thesis*, Queen's University of Belfast, UK, 374 pages.

McKEOGH, E.J., and ERVINE, D.A. (1981). "Air Entrainment rate and Diffusion Pattern of Plunging Liquid Jets." *Chem. Engrg. Science*, Vol. 36, pp. 1161-1172.

McLEAN, F.G., and HANSEN, K.D. (1993). "Roller Compacted Concrete for Embankment Overtopping Protection." *Proc. Spec. Conf. on Geotechnical Practice in Dam Rehabilitation*, ASCE, Raleigh NC, USA, L.R. ANDERSON editor, pp. 188-209.

MAMAK, W. (1964). "River Regulation." *Arkady*, Warsaw, Poland, 380 pages.

MANNING, R. (1890). "On the Flow of Water in Open Channels and Pipes." *Instn of Civil Engineers of Ireland*.

MARCHI, E. (1993). "On the Free-Overfall." *Jl of Hyd. Res.*, IAHR, Vol. 31, No. 6, pp. 777-790.

MARSH, F.B. (1957). "Danger of Fixed Flashboards shown by Flood at Croton Dam." *Civil Engineering*, ASCE, Vol. 64, pp. 408-409.

MASTROPIETRO, M.A. (1968). "Effects of Dam Reaeration on Waste Assimilation Capacities of the Mohawk River." *Proc. 23rd Industrial Waste Conf.*, Engineering Extension Series, No. 132, Part 2, Purdue University, pp. 754-765.

MAULL, D.J., and EAST, L.F. (1963). "Three-Dimensional Flow in Cavities." *Jl of Fluid Mech.*, Vol. 16, pp. 620-632.

MAY, R.W.P. (1987). "Cavitation in Hydraulic Structures : Occurrence and Prevention." *Hydraulics Research Report*, No. SR 79, Wallingford, UK.

MAY, R.W.P., and WILLOUGHBY, I.R. (1991). "Impact Pressures in Plunge Basins Due to Vertical Falling Jets." *Hydraulics Research Report*, No. SR 242, Wallingford, UK.

Metropolitan Water (1980). "Investigation into Spillway Discharge Noise at Avon Dam." *ANCOLD Bulletin*, No. 57, Aug., pp. 31-36.

MICHELS, V. (1966). "Some Recent Spillway Designs." *ANCOLD Bulletin*, No. 19, pp. 15-24.

MICHELS, V., and LOVELY, M. (1953). "Some Prototype Observations of Air Entrained Flow." *Proc. 5th IAHR Congress*, IAHR-ASCE, Minneapolis, USA, pp. 403-414.

MICKELSON, H., and MOORWOOD, R.P. (1984). "The Maranoa River Weir - Design for a Typical Inland Stream." *ANCOLD Bulletin*, No. 67, pp. 37-42. Discussion : No. 67, pp. 42-43.

MIKSIS, M., VANDEN-BROECK, J.M., and KELLER, J.B. (1981). "Axisymmetric Bubble or Drop in a Uniform Flow." *Jl of Fluid Mechanics*, Vol. 108, pp. 89-100.

MILLER, A.P., PRAVDIVETS, Y.P., and SALOV, V.A. (1987). "Earth Overflow Dam." *Gidrotekhnicheskoe Stroitel'stvo*, No. 8, pp. 49-52 (in Russian). (Translated in Hydrotechnical Construction, 1988, Plenum Publ., pp. 503-507).

MONTES, J.S. (1992). "A Potential Flow Solution for the Free Overfall." *Proc. Intn. Civ. Engrs Wat. Marit. & Energy*, Vol. 96, Dec., pp. 259-266.

MOORE, W.L. (1943). "Energy Loss at the Base of a Free Overfall." *Transactions*, ASCE, Vol. 108, p. 1343-1360. Discussion : Vol. 108, pp. 1361-1392.

MORELLI, C. (1971). "The International Gravity Standardization Net 1971 (I.G.S.N.71)." *Bureau Central de l'Association Internationale de Géodésie*, Paris, France.

NAKASONE, H. (1987). "Study of Aeration at Weirs and Cascades." *Jl of Envir. Engrg.*, ASCE, Vol. 113, No. 1, pp. 64-81.

NEEDHAM, J. (1971). "Science and Civilisation in China." *Cambridge University Press*, London, UK, Vol. 4, Part 3, Section 28.

NETZER, E. (1983). "Water Channels and a Royal Estate from the Late Hellenistic Period in the Western Plains of Jericho." *Symp. on Historical Water Development Projects in the Eastern Mediterranian*, Jerusalem, Israël (Mitteilungen, Institut für Wasserbau der Technischen Universität Braunschweig, Germany, No. 82, 1984).

NOETZLI, F.A. (1931). "High Dams - A Symposium - Discussion." *Transactions*, ASCE, Vol. 95, pp. 171-174.

NOORI, B.M.A. (1984). "Form Drag Resistance of Two Dimensional Stepped Steep Open Channels." *Proc. 1st Intl Conf. on Hyd. Design in Water Resources Engineering*, Channels and Channel Control Structures, Southampton, UK, K.V.H. SMITH Ed., Springer-Verlag Publ., pp.

1.133-1.147.

O'LOUGHLIN, E.M., and MACDONALD, E.G. (1964). "Some Roughness Concentration Effects on Boundary Resistance." *Jl La Houille Blanche*, No. 7, pp. 773-783.

OKAMOTO, S., SEO, S., NAKASO, K., and KAWAI, I. (1993). "Turbulent Shear Flow and Heat Transfer over the Repeated Two-Dimensional Square Ribs on Ground Plane." *Jl of Fluids Eng.*, Trans. ASME, Vol. 115, Dec., pp. 631-637.

OLIVIER, H. (1967). "Through and Overflow Rockfill Dams - New Design Techniques." *Proc. Instn. Civil Eng.*, March, 36, pp. 433-471. Discussion, 36, pp. 855-888.

OUTLAND, C.F. (1963). "Man-Made Disaster : the Story of the St. Francis Dam." *Arthur H. Clark Comp.*, Glendale (Cal), USA.

PANDIT, A.B., and DAVIDSON, J.F. (1986). "Bubble Break-up in Turbulent Liquid." *Proc. Intl Conf. Bioreactor Fluid Dynamics*, BHRA Fluid Eng., Cambridge, UK, pp. 109-120.

PARISET, E. (1955). "Etude sur la Vibration des lames Déversantes." ('Study of the Vibration of Free-Falling Nappes.') *Proc. 6th IAHR Congress*, The Hague, The Netherlands, Vol. 3, paper C21, pp. 1-15.

PERRY, A.E., SCHOFIELD, W.H., and JOUBERT, P.N. (1969). "Rough Wall Turbulent Boundary Layers." *Jl of Fluid Mech.*, Vol. 37, Part 2, pp. 383-413.

PETERKA, A.J. (1953). "The Effect of Entrained Air on Cavitation Pitting." *Joint Meeting Paper*, IAHR/ASCE, Minneapolis, Minnesota, Aug. 1953, pp. 507-518.

PETRIKAT, K. (1958). "Vibration Tests on Weirs and Bottom Gates." *Water Power*, pp. 52-57, 99-104, 147-149 & 190-197.

PETRIKAT, K. (1978). "Model Tests on Weirs, Bottom Outlet Gates, Lock Gates and Harbours Moles." *MAN Technical Bulletin*, Nurnberg, Germany.

PEYRAS, L., ROYET, P., and DEGOUTTE, G. (1991). "Ecoulement et Dissipation sur les Déversoirs en Gradins de Gabions." ('Flows and Dissipation of Energy on Gabion Weirs.') *Jl La Houille Blanche*, No. 1, pp. 37-47 (in French).

PEYRAS, L., ROYET, P., and DEGOUTTE, G. (1992). "Flow and Energy Dissipation over Stepped Gabion Weirs." *Jl of Hyd. Engrg.*, ASCE, Vol. 118, No. 5, pp. 707-717.

POST, G., CHERVIER, L., BLANCHET, C., and JOHNSON, G. (1987). "Spillways for Dams." *ICOLD Bulletin*, No. 58, 171 pages.

POWELL, R.W. (1944). "Flow in a Channel of Definite Roughness." *Proceedings*, ASCE, Vol. 70, No. 10, pp. 1521-1544.

PRAVDIVETS, Y.P., and BRAMLEY, M.E. (1989). "Stepped Protection Blocks for Dam Spillways." *Intl Water Power and Dam Construction*, Vol. 41, No. 7, July, pp. 49-56.

Queensland Water Resources (1988). "Construction of Bucca Weir." *Water Resources Commission Report*, Contract No. 2541, Post-Construction Report, April, Brisbane, Australia.

QUINTELA, A.C., CARDOSO, J.L., and MASCARENHAS, J.M. (1987). "Roman Dams in Southern Portugal" *Intl Water Power and Dam Construction*, Vol. 39, No. 5, pp. 38-40 & 70.

RABBEN, S.L., ELS, H., and ROUVE, G. (1983). "Investigation on Flow Aeration at Offsets

Downstream of High-Head Control Structures." *Proc. 20th IAHR Congress*, Moscow, USSR, Vol. 3, pp. 354-360.

RAJARATNAM, N. (1962). "An Experimental Study of Air Entrainment Characteristics of the Hydraulic Jump." *Jl of Instn. Eng. India*, Vol. 42, No. 7, March, pp. 247-273.

RAJARATNAM, N. (1967). "Hydraulic Jumps." *Advances in Hydroscience*, Ed. V.T. CHOW, Academic Press, New York, USA, Vol. 4, pp. 197-280.

RAJARATNAM, N. (1976). "Turbulent Jets." *Elsevier Scientific Publ. Co.*, Development in Water Science, 5, New York, USA.

RAJARATNAM, N. (1990). "Skimming Flow in Stepped Spillways." *Jl of Hyd. Engrg.*, ASCE, Vol. 116, No. 4, pp. 587-591. Discussion : Vol. 118, No. 1, pp. 111-114.

RAJARATNAM, N., and KATOPODIS, C. (1991). "Hydraulics of Steeppass Fishways." *Can. Jl. of Civil Eng.*, Vo. 18, pp. 1024-1032.

RAJARATNAM, N., and MURALIDHAR, D. (1968). "Characteristics of the Rectangular Free Overfall." *Jl of Hyd. Res.*, IAHR, Vol. 6, No. 3, pp. 233-258.

RAND, W. (1955). "Flow Geometry at Straight Drop Spillways." *Proceedings*, ASCE, Vol. 81, No. 791, Sept., pp. 1-13.

RENNER, J. (1973). "Lufteinmischung beim Aufprall eines ebenen Wasserstrahls auf eine Wand." ('Air Entrainment by a Plane Water Jet Impinging on a Wall.') *Ph.D. thesis*, University of Karsruhe, Germany (in German).

RENNER, J. (1975). "Air Entrainment in Surface Rollers." *Proc. Symp. on Design & Operation of Siphons & Siphon Spillways*, BHRA Fluid Eng., London, UK, paper A4, pp. 48-56.

RESCH, F.J., and LEUTHEUSSER, H.J. (1972). "Le Ressaut Hydraulique : mesure de Turbulence dans la Région Diphasique." ('The Hydraulic Jump : Turbulence Measurements in the Two-Phase Flow Region.') *Jl La Houille Blanche*, No. 4, pp. 279-293 (in French).

RESCH, F.J., LEUTHEUSSER, H.J., and ALEMU, S. (1974). "Bubbly Two-Phase Flow in Hydraulic Jump." *Jl of Hyd. Div.*, ASCE, Vol. 100, No. HY1, pp. 137-149.

RHONE, T.J. (1990). "Development opf Hydraulic Structures." *50th Anniversary of the Hydraulics Division 1938-1988*, ASCE, Ed. A.M. ALSAFFAR, New York, USA, pp. 132-147.

RICHTER, J.P. (1939). "The Literary Works of LEONARDO DA VINCI." *Oxford University Press*, London, UK, 2 volumes.

RINDELS, A.J., and GULLIVER, J.S. (1986). "Air-Water Oxygen Transfer at Spillways and Hydraulic Jumps." *Proc. Water Forum '86: World Water Issues in Evolution*, ASCE, New York, USA, pp. 1041-1048.

ROBERTS, P.R. (1977). "Energy Dissipation by Dam Crest Splitters." *The Civil Engineer in South Africa*, Nov., pp. 263-264.

ROBINSON, K.M. (1989). "Hydraulic Stresses on an Overfall Boundary." *Transactions*, ASAE, Vol. 32, No. 4, pp. 1269-1274.

ROLT, L.T.C. (1973). "From Sea to Sea - The Canal du Midi." *Allen Lane*, London, UK, 198 pages.

ROUSE, H. (1936). "Discharge Characteristics of the Free Overfall." *Civil Engineering*, Vol. 6,

April, p. 257.

ROUSE, H. (1938). "Fluid Mechanics for Hydraulic Engineers." *McGraw-Hill*, New York, USA.

ROUSE, H. (1943). "Energy Loss at the Base of a Free Overfall - Discussion." *Transactions*, ASCE, Vol. 108, pp. 1383-1387.

RUSSELL, S.O., and SHEEHAN, G.J. (1974). "Effect of Entrained Air on Cavitation Damage." *Can. Jl of Civil Engrg.*, Vol. 1, pp. 97-107.

RYSKIN, G., and LEAL, L.G. (1984). "Numerical Solution of Free-Boundary Problems in Fluid Mechanics. Part 3. Bubble Deformation in an Axisymmetric Straining Flow." *Jl of Fluid Mechanics*, Vol. 48, pp. 37-43.

SAMS, E.W. (1952). "Experimental Investigation of Average Heat-Transfer and Friction Coefficients for Air Flowing in Circular Tubes Having Square-Thread-Type Roughness." *NACA report*, RM E52D17, USA, 43 pages.

SAYRE, W.W., and ALBERTSON, M.L. (1963). "Roughness Spacing in Rigid Open Channels." *Transactions*, ASCE, Vol. 128, pp. 343-372. Discussion : Vol. 128, pp. 372-427.

SCHEUERLEIN, H. (1973). "The Mechanics of Flow in Steep, Rough Open Channels." *Proc. 15th IAHR Congress*, Istanbul, Turkey, Paper A56, pp. 1-9.

SCHLICHTING, H. (1979). "Boundary Layer Theory." *McGraw-Hill*, New York, USA, 7th edition.

SCHNITTER, N.J. (1991). "Roman Dams and Weirs in the Iberian Peninsula." *8th Intl Symp. on History of Hydraulics*, Merida, Spain (Mitteilungen, Inst. für Wasserbau der Technische Univ. Braunschweig, Germany, Col. 117, 1992, pp. 161-177).

SCHNITTER, N.J. (1994). "A History of Dams : the Useful Pyramids." *Balkema Publ.*, Rotterdam, The Netherlands .

SCHODEK, D.L. (1987). "Landmarks in American Civil Engineering." *MIT Press*, Cambridge MA, USA.

SCHOKLITSCH, A. (1937). "Hydraulic structures." *ASME*, New York, USA, Vol. 2.

SCHRÖDER, R. (1963). "Die turbulente Strömung im freien Wechselsprung." *Mitteilungen*, Institut für Wasserbau und Wasserwirtschaft, T.U. Berlin, Germany, Heft 59.

SCHUYLER, J.D. (1909). "Reservoirs for Irrigation, Water-Power and Domestic Water Supply." *John Wiley & sons*, 2nd edition, New York, USA.

SCHWARTZ, H.I. (1964a). "Projected Nappes Subject ot Harmonic Pressures." *Proc. Instn. Civ. Engrs. (London)*, Vol. 28, pp. 313-36.

SCHWARTZ, H.I. (1964b). "Nappe Oscillation." *Jl of Hyd. Div.*, ASCE, Vol. 90, No. HY6, pp. 129-143. Discussion : Vol. 91, No. HY3, pp. 389-392.

SCHWARTZ, H.I., and NUTT, L.P. (1963). "Projected Nappes Subject to Transverse Pressure." *Jl of Hyd. Div.*, ASCE, Vol. 89, July, pp. 97-104.

SENE, K.J. (1984). "Aspects of Bubbly Two-Phase Flow." *Ph.D. thesis*, Trinity College, Cambridge, UK, Dec..

SENE, K.J. (1988). "Air Entrainment by Plunging Jets." *Chem. Eng. Science*, Vol. 43, Bo. 10, pp. 2615-2623.

SEVIK, M., and PARK, S.H. (1973). "The Splitting of Drops and Bubbles by Turbulent Fluid Flow." *Jl Fluids Engrg.*, Trans. ASME, Vol. 95, pp. 53-60.

SKOGLUND, V.J. (1936). "Effect of Roughness on the Friction Coefficient of a Closed Channel." *Jl of Aeronaut. Sciences*, Vol. 4, Nov., pp. 28-29.

SMITH, H.A. Jr. (1973). "A Detrimental Effect of Dams on Environment : Nitrogen Supersaturation." *Proc. 11th ICOLD Congress*, Madrid, Spain, Q. 40, R. 17, pp. 237-253.

SMITH, N. (1971). "A History of Dams." *The Chaucer Press*, Peter Davies, London, UK.

Soil and Water (1992). *Soil and Water Newsletter*, Council of Agriculture, Executive Yuan, Taiwan, No. 13, Sept..

SORENSEN, R.M. (1985). "Stepped Spillway Hydraulic Model Investigation." *Jl of Hyd. Engrg.*, ASCE, Vol. 111, No. 12, pp. 1461-1472. Discussion : Vol. 113, No. 8, pp. 1095-1097.

SOYER, G. (1992). "Evacuateur de Crues du Barrage de Pinet sur le Tarn." ('Spillway of the Pinet Dam on the Tarn River.') *Jl La Houille Blanche*, No 2/3, pp. 195-196. Discussion, p. 248.

STEIN, A. (1940). "Surveys on the Roman Frontier in Iraq and Trans-Jordan." *The Geographical Journal*, Vol. XCV, pp. 428-438.

STEPHENSON, D. (1979a). "Gabion Energy Dissipators." *Proc. 13th ICOLD Congress*, New Delhi, India, Q. 50, R. 3, pp. 33-43.

STEPHENSON, D. (1979b). "Rockfill in Hydraulic Engineering." *Elsevier Scientific Publ. Comp.*, Amsterdam, The Netherlands.

STEPHENSON, D. (1980). "The Stability of Gabion Weirs." *Intl Water Power and Dam Construction*, Vol. 31, No. 4, pp. 24-28.

STEPHENSON, D. (1988). "Stepped Energy Dissipators." *Proc. Intl Symp. on Hydraulics for High Dams*, IAHR, Beijing, China, pp. 1228-12-35.

STEPHENSON, D. (1991). "Energy Dissipation down Stepped Spillways." *Intl Water Power and Dam Construction*, Vol. 43, No. 9, Sept., pp. 27-30.

STEWART, W.G. (1913). "The Determination of the N in Kutter's Formula for Various Canals, Flumes and Chutes on the Boise Project and Vicinity." *Report on 2nd Annual Conf. on Operating Men*, USBR, Boise, Idaho, USA, Jan., pp. 8-23.

STRAUB, L.G., and ANDERSON, A.G. (1958). "Experiments on Self-Aerated Flow in Open Channels." *Jl of Hyd. Div.*, Proc. ASCE, Vol. 84, No. HY7, paper 1890, pp. 1890-1 to 1890-35.

STRAUB, L.G., and LAMB, O.P. (1953). "Experimental Studies of Air Entrainment in Open-Channel Flows." *Proc. 5th IAHR Congress*, IAHR-ASCE, Minneapolis, USA, pp. 425-437.

STREETER, V.L. (1936). "Frictional Resistance in Artificially Roughened Pipes." *Transactions*, ASCE, Vol. 101, pp. 681-704. Discussion : Vol. 101, pp. 705-713.

STREETER, V.L., and CHU, H. (1949). "Fluid Flow and Heat Transfer in Artificially Roughened Pipes." *Report Project 4918*, Armour Research Foundation, Chicago, USA.

STREETER, V.L., and WYLIE, E.B. (1981). "Fluid Mechanics." *McGraw-Hill*, 1st SI Metric edition, Singapore.

SUN, T.Y., and FAETH, G.M. (1986). "Structure of Turbulent Bubbly Jets-II. Phase Property

Profiles." *Intl Jl of Multiphase Flow*, Vol. 12, No. 1, pp. 115-126.

TEBBUTT, T.H.Y. (1972). "Some Studies on Reaeration in Cascades." *Water Research*, Vol. 6, pp. 297-304.

TEBBUTT, T.H.Y., ESSERY, I.T.S., and RASARATNAM, S.K. (1977). "Reaeration Performance of Stepped Cascades." *Jl Instn. Water. Eng. Sci.*, Vol. 31, No. 4, pp. 285-297.

THANDAVESWARA, B.S. (1974). "Self Aerated Flow Characteristics in Developing Zones and in Hydraulic Jumps." *Ph.D. thesis*, Dept. of Civil Engrg., Indian Institute of Science, Bangalore, India, 399 pages.

THOMAS, H.H. (1976). "The Engineering of Large Dams." *John Wiley & Sons*, London, UK, 2 Volumes.

THOMAS, N.H., AUTON, T.R., SENE, K., and HUNT, J.C.R. (1983). "Entrapment and Transport of Bubbles by transient Large Eddies in Multiphase Turbulent Shear Flows." *Proc. Intl Conf. on Physical Modelling of Multiphase Flow*, BHRA Fluid Engrg., Coventry, UK, pp. 169-184.

THOMPSON, R.C., and HUTCHINSON, R.W. (1929). "The Excavations of the Temple of Nabû at Nineveh." *Archaeologia (London)*, Vol. 79, pp. 103-148.

THOMPSON, S.M., and CAMPBELL, P.L. (1979). "Hydraulics of a Large Channel Paved with Boulders." *Jl of Hyd. Res.*, IAHR, Vol. 17, No. 4, pp. 341-354.

THORNE, C.R., and ZEVENBERGEN, L.W. (1985). "Estimating Velocity in Mountain Rivers." *Jl of Hyd. Engrg.*, ASCE, Vol. 111, No. 4, p. 612-624.

TOMINAGA, A., and NEZU, I. (1991). "Turbulent Structure Past Strip Roughness in Open Channel Flows." *Proc. 24th IAHR Congress*, Madrid, Spain, Vol. C, pp. 42-50.

TOSO, J.W., and BOWERS, C.E. (1988). "Extreme Pressures in Hydraulic-Jump Stilling Basins." *Jl of Hyd. Engrg.*, ASCE, Vol. 114, No. 8, pp. 829-843.

TOWNES, H.W., and SABERSKY, R.H. (1966). "Experiments on the Flow over a Rough Surface." *Intl Jl of Heat and Mass Transfer*, Vol. 9, pp. 729-738.

TOZZI, M.J. (1992). "Caracterização/Comportamento de Escoamentos em Vertedouros com Paramento em Degraus." ('Hydraulics of Stepped Spillways.') *Ph.D. thesis*, University of Sao Paulo, Brazil (in Portuguese).

UTRILLAS, J.L., GAMO, A., and SORIANO, A. (1992). "Reconstruction of the Tous Dam." *Intl Water Power and Dam Construction*, Vol. 44, No. 9, Sept., pp. 55-65.

VAN DE DONK, J. (1981). "Water Aeration with Plunging Jets." *Ph.D. thesis*, TH Delft, The Netherlands, 168 pages.

VAN DE SANDE, E., and SMITH, J.M. (1972). "Eintragen von Luft in eine Flüssigkeit durch eine Wasserstrahl Teil 1 : Strahlen mit geringer Geschwindigkeit." *Chem. Inger. Tech.*, Vol. 44, No. 20, pp. 1177-1183. (in German)

VAN DE SANDE, E., and SMITH, J.M. (1973). "Surface Entrainment of Air by High Velocity Water Jets." *Chem. Eng. Science*, Vol. 28, pp. 1161-1168.

VAN DE SANDE, E., and SMITH, J.M. (1976). "Jet Break-up and Air Entrainment by Low Velocity Turbulent Water Jets." *Chem. Eng. Science*, Vol. 31, pp. 219-224.

VANDINE, D.F. (1985). "Debris Flows and Debris Torrents in Southern Canada Cordillera." *Can Jl of Geotech.*, Vol. 22, pp. 44-68.

VASILIEV, O.F., and BUKREYEV, V.I. (1967). "Statistical Characteristics of Pressure Fluctuations in the region of Hydraulic jump." *Proc. 12th IAHR Congress*, Fort-Collins, USA, Vol. 2, paper B1, pp. 1-8.

VITA-FINZI, C. (1961). "Roman Dams in Tripolitania." *Antiquity*, Vol. 35, pp. 14-20.

VITTAL, N., and POREY, P.D. (1987). "Design of Cascade Stilling Basins for High Dam Spillways." *Jl of Hyd. Engrg.*, ASCE, Vol. 113, No. 2, pp. 225-237.

VITTAL, N., RAGA RAJU, K.G., and GARDE, R.J. (1977). "Resistance of Two Dimensional Triangular Roughness." *Jl of Hyd. Res.*, IAHR, Vol. 15, No. 1, pp. 19-36.

VOLKART, P. (1980). "The Mechanism of Air Bubble Entrainment in Self-Aerated Flow." *Intl Jl of Multiphase Flow*, Vol. 6, pp. 411-423.

WARK, R.J., KERBY, N.E., and MANN, G.B. (1991). "New Victoria Dam Project." *ANCOLD Bulletin*, No. 88, Aug., pp. 14-32.

WARK, R.J., and MANN, G.B. (1992). "Design and Construction Aspects of New Victoria Dam." *Intl Water Power and Dam Construction*, Vol. 44, No. 2, Feb., pp. 24-29.

Water Pollution (1957). "Oxygen Balance in Surface Waters." *Water Pollution Research*, DSIR, UK, pp. 12-21.

Water Power (1992). "Five TBMs Break Records at Svartisen." *Intl Water Power and Dam Construction*, Vol. 44, No. 6, June, pp. 23-27.

WEGMANN, E. (1907). "The Design of the New Croton Dam." *Transactions*, ASCE, Vol. LXVIII, No. 1047, pp. 398-457.

WEGMANN, E. (1911). "The Design and Construction of Dams." *John Wiley & Sons*, New York, USA, 6th edition.

WEI, C.Y., and DE FAZIO, F.G. (1982). "Simulation of Free Jet Trajectories for Design of Aeration Devices on Hydraulic Structures." *Proc. 4th Intl. Conf. on Finite Elements in Water Resources*, Hannover, Germany, June, pp. 17/45-17/55.

WEISS, R.F. (1970). "The Solubility of Nitrogen, Oxygen and Argon in Water and Seawater." *Deep-Sea Research*, Vol. 17, pp. 721-735.

WHITE, M.P. (1943). "Energy Loss at the Base of a Free Overfall - Discussion." *Transactions*, ASCE, Vol. 108, pp. 1361-1364.

WILEY, A.J. (1931). "High Dams - A Symposium - Closure." *Transactions*, ASCE, Vol. 95, pp. 212-217.

WILHELMS, S.C., CLARK, L., WALLACE, J.R., and SMITH, D.R. (1981). "Gas Transfer in Hydraulic Jumps." *Technical Report E-81-10*, US Army Engineer Waterways Experiment Station, CE, Vicksburg Miss., USA.

WILHELMS, S.C., and GULLIVER, J.S. (1989). "Self-Aerating Spillway Flow." *National Conference on Hydraulic Engineering*, ASCE, New Orleans, USA, M.A. PORTS editor, pp. 881-533.

WISNER, P. (1965). "Sur le Rôle du Critère de Froude dans l'Etude de l'Entraînement de l'Air par les Courants à Grande Vitesse." ('On the Role of the Froude Criterion for the Study of Air Entrainment in High Velocity Flows.') *Proc. 11th IAHR Congress*, Leningrad, USSR, paper 1.15 (in French).

WOOD, I.R. (1983). "Uniform Region of Self-Aerated Flow." *Jl Hyd. Eng.*, ASCE, Vol. 109, No. 3, pp. 447-461.

WOOD, I.R. (1984). "Air Entrainment in High Speed Flows." *Proc. Intl. Symp. on Scale Effects in Modelling Hydraulic Structures*, IAHR, Esslingen, Germany, H. KOBUS editor, paper 4.1.

WOOD, I.R. (1985). "Air Water Flows." *Proc. 21st IAHR Congress*, Melbourne, Australia, Keynote address, pp. 18-29.

WOOD, I.R. (1991). "Air Entrainment in Free-Surface Flows." *IAHR Hydraulic Structures Design Manual No. 4*, Hydraulic Design Considerations, Balkema Publ., Rotterdam, The Netherlands, 149 pages.

WOOD, I.R., ACKERS, P., and LOVELESS, J. (1983). "General Method for Critical Point on Spillways." *Jl. of Hyd. Eng.*, ASCE, Vol. 109, No. 2, pp. 308-312.

YAGASAKI, T., and KUZUOKA, T. (1979). "Surface Entrainment of Gas by Plunging Liquid Jet." *Research Report No. 47*, Kogakuin University, Japan, pp. 77-85.

YUAN, Mingshun (1990). "On Numerical Simulations of Aerated Jet Flows about aerators on Spillway." *Proc. 7th Congress of APD-IAHR*, Vol. II, Beijing, China, Nov., pp. 166-170.

APPENDIX A
NAPPE FLOW TRAJECTORY AT A DROP STRUCTURE

A.1 Introduction

In a nappe flow regime, the flow from each step hits the step below as a falling jet, and stepped spillway with nappe flows are analysed as a succession of drop structures. For a horizontal step, a simple expression of the nappe trajectory can be deduced from the motion equation assuming that the velocity of the flow at the brink of the step is nearly horizontal. Application of the momentum equation at the base of the overfall provides most flow properties at the jet impact. Most calculations are developed for aerated nappes. The particular case of un-aerated napped is detailed in sub-section A.3.

A.2 Nappe trajectory of aerated nappe

Considering an aerated nappe, the flow direction at the brink of the step is nearly horizontal. Once the fluid leaves the step (fig. A-1), the horizontal acceleration is zero and the vertical acceleration equals minus the gravity acceleration. The nappe velocity components in the x- and y-directions are :

$$V_x = V_b \tag{A-1a}$$

$$V_y = -g * t \tag{A-1b}$$

where V_b is the flow velocity at the brink of the step and t is the time. The trajectory equations of the nappe centreline are :

$$x = V_b * t \tag{A-2a}$$

$$y = h + \frac{d_b}{2} - \frac{1}{2} * g * t^2 \tag{A-2b}$$

where h is the step height and d_b is the flow depth at the brink.

Drop length

When the fluid leaves the step, the time t', taken to hit the next step is given by :

$$0 = -\frac{1}{2} * g * t'^2 + \left(h + \frac{d_b}{2} \right) \tag{A-3}$$

The length of the drop L_d equals :

$$L_d = V_b * t' \tag{A-4}$$

It yields :

Fig. A-1 - Nappe flow

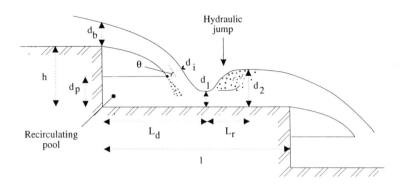

$$\frac{L_d}{h} = \left(\frac{d_c}{h}\right)^{3/2} * \sqrt{\frac{h}{d_b}} * \sqrt{1 + 2 * \frac{h}{d_b}} \qquad (A-5)$$

Impact flow conditions

When the fluid leaves the step, the time t'' taken to reach the pool free-surface is given by :

$$d_p = -\frac{1}{2} * g * t''^2 + \left(h + \frac{d_b}{2}\right) \qquad (A-6)$$

where d_p is the height of water in the pool behind the overfalling jet. The nappe thickness d_i and the flow velocity V_i at the intersection of the falling nappe with the receiving pool are then :

$$\frac{d_i}{d_c} = \left(\left(\frac{d_c}{d_b}\right)^2 + 2 * \frac{h + \frac{d_b}{2} - d_p}{d_c}\right)^{-1/2} \qquad (A-7)$$

$$\frac{V_i}{V_c} = \sqrt{\left(\frac{d_c}{d_b}\right)^2 + 2 * \frac{h + \frac{d_b}{2} - d_p}{d_c}} \qquad (A-8)$$

The angle of the falling nappe with the horizontal (fig. A-1) is given by :

$$\tan\theta = \sqrt{2} * \sqrt{\frac{d_b}{d_c}} * \sqrt{\frac{h + \frac{d_b}{2} - d_p}{d_c}} \qquad (A-9)$$

Pool height

The pool of water behind the overfalling jet (fig. A-1) is important as its weight provides the force parallel to the step surface which is required to change the jet from an angle θ to the step bottom to parallel to the horizontal face of the step. For a ventilated nappe, the momentum equation resolved along the step surface is :

$$\frac{1}{2} * \rho_w * g * d_p^2 - \frac{1}{2} * \rho_w * g * d_1^2 = \rho_w * q_w * (V_1 - V_i * \cos\theta) \tag{A-10}$$

assuming that the edges of the jet fluid have not disintegrated into spray and neglecting the shear forces on the surfaces. Assuming that the velocity entering the control volume is approximately the same as leaving it, it yields :

$$\frac{d_p}{d_i} = \sqrt{1 + 2 * \frac{V_i^2}{g * d_i} * (1 - \cos\theta)} \tag{A-11}$$

Flow depth downstream of the jet impact

WHITE (1943) developed the momentum equation at the base of the overfall to estimate the energy loss. His results gave the flow depth downstream of the jet impact (fig. A-1) :

$$\frac{d_1}{d_c} = \frac{\sqrt{2}}{\frac{1.5}{\sqrt{2}} + \sqrt{\frac{h}{d_c} + \frac{3}{2}}} \tag{A-12}$$

Note that the flow immediately downstream of the jet impact is supercritical : i.e., $d_1/d_c < 1$.

Flow depth downstream of the hydraulic jump

The flow conditions downstream of the hydraulic jump are deduced from the momentum and continuity equations. For a horizontal step, constant channel width, and neglecting the bed and wall friction, it yields (STREETER and WYLIE 1981) :

$$\frac{d_2}{d_1} = \frac{1}{2} * \left(\sqrt{1 + 8 * \left(\frac{d_c}{d_1}\right)^3} - 1 \right) \tag{A-13}$$

where the subscripts 1 and 2 refer respectively to the upstream and downstream flow conditions of the hydraulic jump (fig. A-1).

<u>Discussion</u>

Equations (A-5), (A-7), (A-9), (A-11), (A-12) and (A-13) form a system of six (6) non-linear equations in terms of the drop length L_d, the impact flow conditions d_i and θ, the pool height d_p, the flow depths d_1 and d_2.

Assuming that the flow upstream of the step edge is subcritical and that $d_b/d_c = 0.715$ (ROUSE 1936), the author has solved numerically the system of equations. The results are best fitted by the following correlations :

$$\frac{L_d}{h} = 2.171 * \left(\frac{d_c}{h}\right)^{0.525} \tag{A-14}$$

$$\frac{d_i}{h} = 0.688 * \left(\frac{d_c}{h}\right)^{1.483} \tag{A-15}$$

$$\tan\theta = 0.855 * \left(\frac{d_c}{h}\right)^{-0.582} \tag{A-16}$$

$$\frac{d_p}{h} = 0.998 * \left(\frac{d_c}{h}\right)^{0.675} \tag{A-17}$$

$$\frac{d_1}{h} = 0.625 * \left(\frac{d_c}{h}\right)^{1.326} \tag{A-18}$$

$$\frac{d_2}{h} = 1.565 * \left(\frac{d_c}{h}\right)^{0.809} \tag{A-19}$$

Practical calculations of the nappe geometry

For aerated nappes, table A-1 reviews nappe geometry calculations and experimental observations. All results were obtained in rectangular channels with horizontal overfall crest.

For engineering applications, the flow depth at the brink d_b is computed as ROUSE (1936) :

$$d_b = 0.715 * d_c \tag{A-20}$$

The correlations of RAND (1955) were successfully verified with experimental data. They are commonly used to estimate the drop length, pool height and the flow depths d_1 and d_2 :

$$\frac{L_d}{h} = 4.30 * \left(\frac{d_c}{h}\right)^{0.81} \tag{A21}$$

$$\frac{d_p}{h} = \left(\frac{d_c}{h}\right)^{0.66} \tag{A-22}$$

$$\frac{d_1}{h} = 0.54 * \left(\frac{d_c}{h}\right)^{1.275} \tag{A-23}$$

$$\frac{d_2}{h} = 1.66 * \left(\frac{d_c}{h}\right)^{0.81} \tag{A-24}$$

Using RAND's (1955) results, the impact flow conditions (eq. (A-7) to (A-9)) can be correlated as :

$$\frac{d_i}{d_c} = 0.687 * \left(\frac{d_c}{h}\right)^{0.483} \tag{A-25}$$

$$\frac{V_i}{V_c} = 1.455 * \left(\frac{d_c}{h}\right)^{-0.483} \tag{A-26}$$

$$\tan\theta = 0.838 * \left(\frac{d_c}{h}\right)^{-0.586} \tag{A-27}$$

On stepped channels with nappe flow regime, the experiments of PEYRAS et al. (1991) indicated that the correlations of ROUSE (1936) and RAND (1955) are reasonably accurate for nappe flows with fully-developed developed jumps. For nappe flows with partially-developed hydraulic jump, the correlations of RAND (1955) (i.e. eq. (A-21) to (A-23)) provide reasonable estimate of the flow characteristics at the jet impact (PEYRAS et al. 1991).

Table A-1 - Calculations of the nappe geometry for aerated nappes

Variable (1)	Formula (2)	Ref. (3)	Remarks (4)
d_b [a]	$d_b = 0.715 * d_c$	[RO]	Experimental data.
	$d_b = 0.65 * d_c$	[CR]	Analytical solution.
	$d_b = 0.781 * d_c$	[BA]	Model data. $d_c < 0.124$ m. $W = 0.46$ m.
	$d_b = 0.714 * d_c$	[KR]	Model data. $0.021 < d_c < 0.068$ m. $W = 0.5$ m.
	$0.72 < d_b/d_c < 0.93$	[GI]	Model data. $0.075 < d_c/h < 0.45$.
	$d_b = 0.76 * d_c$	[FE]	Model data. $0.15 < d_c/h < 0.93$.
L_d	$\dfrac{L_d}{h} = 2.17 * \left(\dfrac{d_c}{h}\right)^{0.525}$	[RA]	Model data. $0.045 < d_c/h < 1$.
	$\dfrac{L_d}{h} = 4.30 * \left(\dfrac{d_c}{h}\right)^{0.81}$		Best fit of equation (A-5).
d_i	$\dfrac{d_i}{h} = 0.688 * \left(\dfrac{d_c}{h}\right)^{1.483}$		Best fit of equation (A-7).
$\tan\theta$	$\tan\theta = 0.855 * \left(\dfrac{d_c}{h}\right)^{-0.582}$		Best fit of equation (A-9).
d_p	$\dfrac{d_p}{d_c} = \sqrt{\left(\dfrac{d_1}{d_c}\right)^2 + 2 * \dfrac{d_c}{d_1} - 3}$	[MO]	Analytical solution.
	$\dfrac{d_p}{h} = \left(\dfrac{d_c}{h}\right)^{0.66}$	[RA]	Model data. $0.045 < d_c/h < 1$.
	$\dfrac{d_p}{h} = 1.067 * \left(\dfrac{d_c}{h} - 0.0016\right)^{0.697}$	[GI]	Model data. $0.075 < d_c/h < 0.45$.
	$\dfrac{d_p}{h} = 0.998 * \left(\dfrac{d_c}{h}\right)^{0.675}$		Best fit of equation (A-11)
d_1	$\dfrac{d_1}{d_c} = \dfrac{\sqrt{2}}{\dfrac{1.5}{\sqrt{2}} + \sqrt{\dfrac{h}{d_c} + \dfrac{3}{2}}}$	[WH]	Analytical solution.
	$\dfrac{d_1}{h} = 0.54 * \left(\dfrac{d_c}{h}\right)^{1.275}$	[RA]	Model data. $0.045 < d_c/h < 1$.
	$\dfrac{d_1}{h} = 0.625 * \left(\dfrac{d_c}{h}\right)^{1.326}$		Best fit of equation (A-12).
d_2	$\dfrac{d_2}{d_1} = \dfrac{1}{2} * \left(\sqrt{1 + 8 * Fr_1{}^2} - 1\right)$		Analytical solution.
	$\dfrac{d_2}{h} = 1.66 * \left(\dfrac{d_c}{h}\right)^{0.81}$	[RA]	Model data. $0.045 < d_c/h < 1$.
	$\dfrac{d_2}{h} = 1.565 * \left(\dfrac{d_c}{h}\right)^{0.809}$		Best fit of equation (A-13).

Notes :

[a] :   subcritical flow in rectangular channel

[BA] BAUER and GRAF (1971); [CR] CRAYA (1948); [FE] FERRO (1992); [GI] GILL (1979); [KR]

KRAIJENHOFF and DOMMERHOLT (1977); [MO] MOORE (1943); [RA] RAND (1955); [RO] ROUSE (1936); [WH] WHITE (1943).

A.3 Nappe trajectory of un-aerated nappe

Considering an un-aerated nappe, the pressure in the cavity between the nappe and the step is $P_{atm} - \Delta P$, where P_{atm} is the atmospheric pressure. At the brink of the step, the vertical acceleration of the flow Γ_y equals :

$$\Gamma_y = -g * \left(1 + \frac{\Delta P}{\rho_w * g * d_b} \right) \tag{A-28}$$

The complete jet trajectory can be computed as a function of the flow properties at the brink of the step and the cavity subpressure using the finite element method (WEI and DE FAZIO 1982), analytical methods (SCHWARTZ and NUTT 1963; CHANSON 1988) or a numerical method (YUAN 1990).

CHANSON (1988) developed a two dimensional jet calculation based on the continuity and momentum equations. In the general case, the solution of the trajectory equations, is :

$$x' = \frac{Fr_b^2}{P_{Nb}} * x \sqrt{-\frac{P_{Nb}^2}{Fr_b^4} * y'^2 + 2 * \frac{1 + P_{Nb} * \cos\theta_b}{Fr_b^2} * y' + (\sin\theta_b)^2}$$

$$+ \frac{Fr_b^2}{P_{Nb}^2} * ArcSin\left(\frac{(1 + P_{Nb} * \cos\theta_b) - \frac{P_{Nb}^2}{Fr_b^2} * y'}{\sqrt{1 + 2 * P_{Nb}^2 * \cos\theta_b + P_{Nb}^2}} \right)$$

$$- \frac{Fr_b^2}{P_{Nb}} * \left(\sin\theta_b + \frac{1}{P_{Nb}} * ArcSin\left(\frac{1 + P_{Nb} * \cos\theta_b}{\sqrt{1 + 2 * P_{Nb} * \cos\theta_b + P_{Nb}^2}} \right) \right) \tag{A-29}$$

assuming that ΔP is non zero, and where $y' = y_1/d_b$, $x' = x_1/d_b$, $P_{Nb} = \Delta P/(\rho_w * g * d_b)$, $Fr_b = V_b/\sqrt{g * d_b}$ and θ_b is the initial angle of the streamlines (fig. A-2).

If the pressure difference across the jet is zero (i.e. $\Delta P = 0$), the analytical solution is the same as for the method of SCHWARTZ and NUTT (1963) :

$$x' = Fr_b^2 * \cos\theta_b * \left(\sqrt{\frac{2}{Fr_b^2} * y' + (\sin\theta_b)^2} - \sin\theta_b \right) \tag{A-30}$$

Equations (A-20) and (A-30) apply to both subcritical and supercritical flow upstream of the step edge. The effects of the initial velocity distribution can be taken into account in these calculations. Complete calculations were detailed in CHANSON (1988, 1992).

Fig. A-2 - Jet trajectory

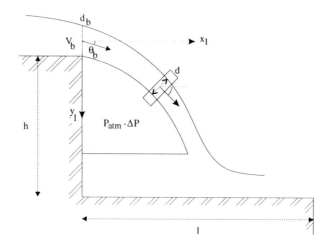

A.4 Frequency of nappe oscillations

Thin sheets of free-falling water are subjected to fluttering instabilities. The instability is related to the shear flow instability of surfaces separating fluids moving at different velocities (CASPERSON 1993) : they are referred as Kelvin-Helmholtz instabilities.

For a nappe leaving a horizontal step, the possible frequencies of nappe oscillation can be determined as (SCHWARTZ 1964b, CASPERSON 1993) :

$$\frac{F * d_b}{\sqrt{g * d_b}} = \frac{I + \frac{1}{4}}{Fr_b * \left(-1 + \sqrt{1 + \frac{2}{Fr_b^2} * \frac{h}{d_b}}\right)} \qquad (A\text{-}31)$$

where I is an integer : I = 1, 2, 3, 4 or more. I (or [I + 1/4]) is approximately the number of wavelengths in the nappe. Estimating the flow depth at the brink as ROUSE (1936), it yields :

$$F * \sqrt{\frac{d_c}{g}} = 0.715 * \left(\frac{I + \frac{1}{4}}{-1 + \sqrt{1 + 1.022 * \frac{h}{d_c}}}\right) \qquad (A\text{-}32)$$

APPENDIX B
PHYSICAL AND CHEMICAL PROPERTIES OF FLUIDS

B.1 Acceleration of gravity

Standard acceleration of gravity

The standard acceleration of gravity is :

$$g = 9.80665 \ m/s^2 \tag{B-1}$$

This "standard" gravity corresponds roughly to that at sea level and 45-degree latitude. The gravitational acceleration varies with latitude and elevation owing to the form and rotation of the earth. ROUSE (1938) proposed the empirical correlation :

$$g = 9.806056 - 0.025027 * \cos(2 * Latitude) - 3E\text{-}6 * z \quad (in \ m/s^2) \tag{B-2}$$

where z is the altitude positive upwards (in metres) with the sea level as the origin.

Geographic altitude z m (1)	Standard acceleration of gravity g m/s² (2)
-1000	9.810
0	9.807
1000	9.804
2000	9.801
3000	9.797
4000	9.794
5000	9.791
6000	9.788
7000	9.785
10000	9.776

Absolute gravity values

Location (1)	g m/s² (2)	Location (1)	g m/s² (2)	Location (1)	g m/s² (2)
Addis Ababa, Ethiopia	9.7743	Helsinki, Finland	9.81090	Québec, Canada	9.80726
Algiers, Algeria	9.79896	Kuala Lumpur, Malaysia	9.78034	Quito, Ecuador	9.7726
Anchorage, USA	9.81925	La Paz, Bolivia	9.7745	Sapporo, Japan	9.80476
Ankara, Turkey	9.79925	Lisbon, Portugal	9.8007	Reykjavik, Iceland	9.82265
Aswan, Egypt	9.78854	Manilla, Philippines	9.78382	Taipei, Taiwan	9.7895
Bangkok, Thailand	9.7830	Mexico city, Mexico	9.77927	Teheran, Iran	9.7939
Bogota, Colombia	9.7739	Nairobi, Kenya	9.77526	Thule, Greenland	9.82914
Brisbane, Australia	9.794	New Delhi, India	9.79122	Tokyo, Japan	9.79787
Buenos Aires, Argentina	9.7949	Paris, France	9.80926	Vancouver, Canada	9.80921
Christchurch, N.Z.	9.8050	Perth, Australia	9.794	Ushuaia, Argentina	9.81465
Denver, USA	9.79598	Port-Moresby, P.N.G.	9.782		
Guatemala, Guatemala	9.77967	Pretoria, South Africa	9.78615		

Reference :   MORELLI (1971)

B.2 Properties of water

Temperature	Density	Dynamic viscosity	Surface tension	Vapour pressure
	ρ_w	μ_w	σ	P_v
Celsius	kg/m^3	Pa.s	N/m	Pa
(1)	(2)	(3)	(4)	(5)
0	999.9	1.792E-3	0.0762	0.6E+3
5	1000.0	1.519E-3	0.0754	0.9E+3
10	999.7	1.308E-3	0.0748	1.2E+3
15	999.1	1.140E-3	0.0741	1.7E+3
20	998.2	1.005E-3	0.0736	2.5E+3
25	997.1	0.894E-3	0.0726	3.2E+3
30	995.7	0.801E-3	0.0718	4.3E+3
35	994.1	0.723E-3	0.0710	5.7E+3
40	992.2	0.656E-3	0.0701	7.5E+3

Reference :   STREETER and WYLIE (1981)

B.3 Salinity and chlorinity

The salinity characterises the amount of dissolved salts in the water : e.g, carbonates, bromides, organic matter. The definition of the salinity is based on the electrical conductivity of water relative to a specified solution of KCl and H_2O (BOWIE et al. 1985). The scale is dimensionless but the salinity is defined often in parts per thousand (ppt).

The chlorinity Chl is defined in relation to salinity Sal as :

$$Sal = 1.80655 * Chl \qquad (B-3)$$

References :   BOWIE et al. (1985)

B.4 Solubility of nitrogen, oxygen and argon in water

Solubility of oxygen

Temperature	C_s(Pstd) Oxygen		
	[Chl = 0]	[Chl = 10 ppt]	[Chl = 20 ppt]
Celsius	kg/m^3	kg/m^3	kg/m^3
(1)	(2)	(3)	(4)
0	14.621E-3	12.388E-3	11.355E-3
5	12.770E-3	11.320E-3	10.031E-3
10	11.238E-3	10.058E-3	8.959E-3
15	10.084E-3	9.027E-3	8.079E-3
20	9.0982E-3	8.174E-3	7.346E-3
25	8.263E-3	7.457E-3	6.728E-3
30	7.559E-3	5.845E-3	6.197E-3
35	6.950E-3	6.314E-3	5.734E-3
40	6.412E-3	5.842E-3	5.321E-3

References :   BOWIE et al. (1985)

The solubility of oxygen in water at equilibrium with water saturated air C_S(Pstd) at standard pressure (i.e. 1.0 atm) is calculated as:

$$LN(1000 * C_S(Pstd)) = -139.34411 + \frac{1.575701E+5}{TC}$$

$$- \frac{6.642308E+7}{TC^2} + \frac{1.243800E+3}{TC^3} - \frac{8.621949E+11}{TC^4}$$

$$- Chl * \left(3.1929E-2 - \frac{19.428}{TC} + \frac{3.8673E+3}{TC^2}\right) \tag{B-4}$$

where C_S(Pstd) is the solubility of oxygen at standard pressure in kg/m^3, TC is the temperature in Celsius and Chl is the chlorinity in ppt. The saturation concentration of dissolved oxygen at non-standard pressure is :

$$C_S(P) = C_S(Pstd) * P * \left(\frac{\left(1 - \frac{P_v}{P}\right) * (1 - Teta*P)}{(1 - P_v) * (1 - Teta)}\right) \tag{B-5}$$

where P is the absolute pressure in atm (within 0 to 2 atm), P_v is the partial pressure of water vapour in atm and TK is the temperature in Kelvin. In equation (B-5), the expression of Teta is :

$$Teta = 0.000975 - 1.426E-5 * TC + 6.436E-8 * TC^2 \tag{B-6}$$

The partial pressure of water vapour may be computed as :

$$LN(P_v) = 11.8571 - \frac{3840.70}{TK} - \frac{216961}{TK^2} \tag{B-7}$$

where P_v is in atm.

References :   APHA (1985,1989), BOWIE et al. (1985)

Volumetric solubility of nitrogen, oxygen and argon

WEISS (1970) proposed an expression of the volumetric solubility of nitrogen, oxygen and argon at one atmosphere total pressure (i.e. standard pressure) :

$$LN(C_S(Pstd)) = A1 + A2 * \frac{100}{TK} + A3 * LN\left(\frac{TK}{100}\right) + A4 * \frac{TK}{100}$$

$$+ Sal * \left(B1 + B2 * \frac{TK}{100} + B3 * \left(\frac{TK}{100}\right)^2\right) \tag{B-8}$$

where C_S(Pstd) is the solubility in ml/kg, TK is the temperature in Kelvin and Sal is the salinity in ppt. The constants A1, A2, A3, A4, B1, B2 and B3 are summarised in the next table for nitrogen (N_2), oxygen (O_2) and argon (Ar).

Gas	A1	A2	A3	A4	B1	B2	B3
(1)	(2)	(3)	(4)	(5)	(6)	(7)	(8)
Nitrogen	-172.4965	248.4262	143.0738	-21.7120	-0.049781	0.025018	-0.0034861
Oxygen	-173.4292	249.6339	143.3483	-21.8492	-0.033096	0.014259	-0.0017000
Argon	-173.5146	245.4510	141.8222	-21.8020	-0.034474	0.014934	-0.0017729

Reference :   WEISS (1970)

B.5 Diffusion coefficients

The gas-liquid diffusivity of oxygen and nitrogen in water is :

Temperature	$D(O_2)$	$D(N_2)$
Celsius	Oxygen m^2/s	Nitrogen m^2/s
(1)	(2)	(3)
10	1.54E-9	1.29E-9
25	2.20E-9	2.01E-9
40	3.33E-9	2.83E-9
55	4.50E-9	3.80E-9

Reference :   FERRELL and HIMMELBLAU (1967)

The data of FERRELL and HIMMELBLAU (1967) can be correlated as :

$$D(O_2) = 1.16793E\text{-}27 * TK^{7.3892} \qquad\qquad\qquad \text{Oxygen (B-9)}$$

$$D(N_2) = 5.567E\text{-}11 * TK - 1.453E\text{-}8 \qquad\qquad\qquad \text{Nitrogen (B-10)}$$

where D is the molecular diffusivity in m^2/s and TK is the temperature in Kelvin.

B.6 Coefficient of mass transfer K_L

KAWASE and MOO-YOUNG (1992) reviewed some correlations for the mass transfer coefficient (K_L) calculations in turbulent gas-liquid flows. They showed that K_L is almost constant regardless of the bubble size and the flow situations. The transfer coefficient of gas bubbles affected by surface active impurities can be correlated as :

$$K_L = 0.28 * D_{gas}^{2/3} * \left(\frac{\mu_w}{\rho_w}\right)^{-1/3} * \sqrt[3]{g} \qquad\qquad (d_{ab} < 0.25 \text{ mm}) \ \text{(B-11)}$$

$$K_L = 0.47 * \sqrt{D_{gas}} * \left(\frac{\mu_w}{\rho_w}\right)^{-1/6} * \sqrt[3]{g} \qquad\qquad (d_{ab} > 0.25 \text{ mm}) \ \text{(B-12)}$$

where D_{gas} is the molecular diffusivity, μ_w and ρ_w are the dynamic viscosity and density of the liquid, d_{ab} is the gas bubble diameter and g is the gravity constant. All variables are expressed in SI units. Equations (B-11) and (B-12) were compared successfully with more than a dozen of experimental studies.

Typical values for oxygen and nitrogen transfer in water are summarised below.

Temperature	$K_L (O_2)$ in water		$K_L (N_2)$ in water	
	Small bubbles m/s	Large bubbles m/s	Small bubbles m/s	Large bubbles m/s
Celsius	Eq. (B-11)	Eq. (B-12)	Eq. (B-11)	Eq. (B-12)
(1)	(2)	(3)	(4)	(5)
5	6.35E-5	3.44E-4	5.06E-5	2.90E-4
10	7.31E-5	3.77E-4	6.49E-5	3.45E-4
15	8.32E-5	4.11E-4	7.55E-5	3.83E-4
20	9.44E-5	4.47E-4	8.82E-5	4.25E-4
25	1.05E-4	4.80E-4	9.90E-5	4.59E-4

Note : calculations performed assuming $g = 9.807$ m/s^2.

APPENDIX C
BUBBLE SIZE CALCULATIONS AT INCLINED PLANE LIQUID JETS

C.1 Introduction

In turbulent shear flows, the maximum bubble size d_m is determined by the balance between capillary forces and the inertial force caused by velocity changes over distances of the order of the bubbles size. Dimensional analysis yields :

$$(We)_c = \frac{\rho_w * v'^2 * d_m}{2 * \sigma} \tag{C-1}$$

where v'^2 is the spatial average value of the square of the velocity differences over a distance equal to d_m and $(We)_c$ is a constant near unity (table 6-7). An assumption is that the term v'^2 is of the order of magnitude of

$$v'^2 \sim \left(\frac{dV}{dy} * d_m\right)^2 \tag{C-2}$$

where V is the local velocity and y is the direction normal to the streamline. It yields :

$$d_m \sim \sqrt[3]{\frac{2 * \sigma * (We)_c}{\rho_w * \left(\frac{dV}{dy}\right)^2}} \tag{C-3}$$

Considering an inclined plunging jet (fig. C-1), the developing flow region is characterised by an upper and a lower shear layer region. Assuming that the air bubbles are entrained tangentially to the liquid jet interfaces, the entrained bubbles are subjected to different buoyancy effects. At the outer (upper) shear layer (fig. C-1(B)), the buoyancy tends to move the bubbles away from the region of high velocity gradient and shear stress. In the inner (lower) shear region, the buoyancy brings the bubbles within regions of higher shear stress. As a result, it is expected that larger air bubbles are entrained at the outer shear layer compared with the lower shear layer (fig. C-1(C)). In the shear regions, neglecting the viscous force and the gravity effect, the solution of the momentum and continuity equation yields to (RAJARATNAM 1976, SCHLICHTING 1979) :

$$\frac{V}{V_o} = \Phi_1(x, y) \tag{C-4}$$

where V_o is the initial jet velocity and x is the distance along the jet centreline

C.2 Upper (outer) shear layer

For a high-velocity jet (i.e. $V_o >> u_r$), the buoyancy effects are small in the jet flow direction. But in the direction normal to the streamlines, the buoyancy force induces a lateral velocity component approximately equal to $u_r * \cos\theta$, where u_r is the bubble rise velocity.
Assuming that the bubbles follow a path diverging from the jet centreline by $u_r * \cos\theta / V_o$ (fig. C-

1(B)), the maximum size of the entrained bubbles is fixed by the maximum velocity gradient $(dV/dy)_{max}$ encountered along their path :

$$\left(\frac{dV}{dy}\right)_{max} = (\Phi_2(V_0, d_{ab}))_{along\ path} \qquad (C-5)$$

and the equation of the bubble path is :

$$y = \frac{d_{ab}}{2} + \frac{u_r * \cos\theta}{V_0} * \left(x - \frac{d_{ab}}{2 * \tan\theta}\right) \qquad (C-6)$$

Fig. C-1 - Idealised air bubble entrainment by an inclined plunging jet

(A) Definitions

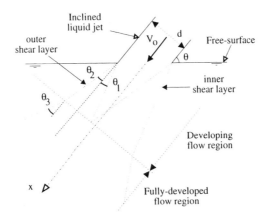

(B) Idealised air bubble entrainment at the upper (outer) shear layer

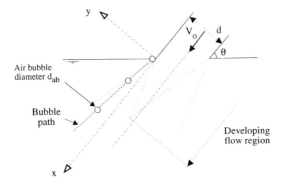

Fig. C-1 - Idealised air bubble entrainment by an inclined plunging jet

(C) Idealised air bubble entrainment at the lower (inner) shear layer

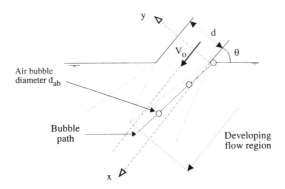

where d_{ab} is the size of the entrained air bubbles. If d_{ab} is larger than the maximum bubble size satisfying equation (C-1), the air bubble is broken up into bubbles of smaller sizes as they are entrained in the regions of high turbulent shear.

The maximum bubble size d_m is the solution of the non-linear system of equations (C-3) and (C-5). Note that, for $d_{ab} < d_m$, the air bubbles are entrained without breaking up.

Application

For a two-dimensional plane jet GOERTLER (1942) solved the continuity and momentum equations assuming that the turbulent kinematic viscosity v_t is estimated as :

$$v_t = \frac{V_o}{4 * K_3^2} * x \qquad (C-7)$$

where K_3 is a constant. In the developing flow region, the velocity follows a Gaussian distribution (GOERTLER 1942, SCHLICHTING 1979) :

$$V = \frac{V_o}{2} * \left(1 + \frac{2}{\sqrt{\pi}} * \int_0^{\xi} \exp(-z^2) * dz \right) \qquad (C-8)$$

where $\xi = K_3*(y - y_{50})/x$ and y_{50} is the location where $V = V_o/2$. The velocity gradient dV/dy equals :

$$\frac{dV}{dy} = \frac{V_o * K_3}{\sqrt{\pi} * x} * \exp\left(-\left(K_3 * \frac{y - y_{50}}{x} \right)^2 \right) \qquad (C-9)$$

Along the bubble path (eq. (C-6)) and assuming that y_{50} is estimated as : $y_{50} = K_4*x$, the velocity gradient dV/dy is maximum at the location x_o such as :

$$x_o = \frac{K_3 * d_{ab}}{2} * \left(1 - \frac{u_r}{V_o} * \frac{\cos\theta}{\tan\theta}\right) *$$
$$* \left(\sqrt{2 + K_3^2 * \left(\frac{u_r}{V_o} * \cos\theta - K_4\right)^2} + K_3 * \left(\frac{u_r}{V_o} * \cos\theta - K_4\right)\right) \quad \text{(C-10)}$$

For an inclined plane jet, the maximum bubble size of air bubbles entrained along the path (eq. (C-6) is obtained by replacing equations (C-10) and (C-9) in equation (C-3).

For the data of LIEPMANN and LAUFER (1947), RAJARATNAM (1976) estimated that $K_3 = 11$, $\theta_1 = 4.8$ degrees, $\theta_2 = 9.5$ degrees (fig. C-1(A)) and $K_4 = 0.041$. With these values of K_3 and K_4, and assuming $u_r = 0.2$ m/s, the author has solved numerically equation (C-3). The result is best correlated by :

$$d_m = (0.05951 + 0.45462 * (\cos\theta)^{2.0503}) * \sqrt[3]{(We)_c}$$
$$* d_{ab}^{2/3} * V_o^{(-2/3 - 1.4931*(\cos\theta)^{1.3853})} \quad \text{(C-11)}$$

where d_m and d_{ab} are in metres, V_o is in m/s and θ is ranging from 90 down to 30 degrees.

Equation (C-11) satisfies the common sense that the maximum bubble size decreases with increasing initial jet velocity. Further, for a given jet velocity, the largest air bubbles are entrained into regions of smaller velocity gradients than small bubbles. As a result the maximum bubble size increases with the initial bubble size. Note that the maximum bubble size is independent of the jet thickness.

A limiting case of equation (C-11) is obtained for $d_m = d_{ab} = (d_m)_*$ where $(d_m)_*$ is the minimum diameter of bubbles that would split in the shear layer :

$$(d_m)_* = (We)_c$$
$$* \left((0.05951 + 0.45462 * (\cos\theta)^{2.0503}) * V_o^{(-2/3 - 1.4931*(\cos\theta)^{1.3853})}\right)^3 \quad \text{(C-12)}$$

Entrained air bubbles of diameter $d_{ab} < (d_m)_*$ are not broken up.

C.3 Lower (inner) shear layer

In the inner shear layer (fig. C-1(C)), the buoyancy force induces a lateral velocity component approximately equal to $u_r*\cos\theta$. With the same reasoning as in paragraph C.2, the maximum size of entrained bubbles is determined by the maximum velocity gradient $(dV/dy)_{max}$ encountered along their path :

$$y = -\frac{d_{ab}}{2} + \frac{u_r * \cos\theta}{V_o} * \left(x + \frac{d_{ab}}{2 * \tan\theta}\right) \quad \text{(C-13)}$$

Along the bubble path (eq. (C-13)) and assuming that y_{50} is estimated as : $y_{50} = - K_4*x$, the velocity gradient dV/dy is maximum at the location x_o which satisfies :

$$x_o = \frac{K_3 * d_{ab}}{2} * \left(1 - \frac{u_r}{V_o} * \frac{\cos\theta}{\tan\theta}\right) *$$

$$* \left(\sqrt{2 + K_3^2 * \left(\frac{u_r}{V_o} * \cos\theta + K_4 \right)^2} - K_3 * \left(\frac{u_r}{V_o} * \cos\theta + K_4 \right) \right) \quad (C\text{-}14)$$

For an inclined plane jet, the maximum bubble size of air bubbles entrained along the path (eq. (C-13) is obtained by replacing equations (C-14) and (C-9) into equation (C-3). Assuming $K_3 = 11$, $\theta_1 = 4.8$ degrees, $\theta_2 = 9.5$ degrees, $K_4 = 0.041$ and $u_r = 0.2$ m/s, the solution is best correlated by :

$$d_m = (0.05951 - 0.04069 * (\cos\theta)^{0.6896}) * \sqrt[3]{(We)_c}$$
$$* d_{ab}^{2/3} * V_o^{(-2/3 + 0.5075*(\cos\theta)^{0.8732})} \quad (C\text{-}15)$$

where d_m and d_{ab} are in metres, V_o is in m/s and θ is ranging from 90 down to 30 degrees. The limiting case of equation (C-15) (i.e. $d_m = d_{ab} = (d_m)_*$) is the minimum diameter of bubbles that would split in the shear layer :

$$(d_m)_* = (We)_c$$
$$* \left((0.05951 - 0.04069 * (\cos\theta)^{0.6896}) * V_o^{(-2/3 + 0.5075*(\cos\theta)^{0.8732})} \right)^3 \quad (C\text{-}16)$$

C.4 Bubble rise velocity

Small air bubbles (i.e. $d_{ab} < 1$ mm) act as rigid spheres. Surface tension imposes the shape of small bubbles but the motion of these bubbles is dominated by the balance between the viscous drag force and the buoyant force. For very small bubbles (i.e. $d_{ab} < 0.1$ mm) the bubble rise velocity u_r is given by the Stokes' law (STREETER and WYLIE 1981) :

$$u_r = \frac{2}{9} * \frac{g * (\rho_w - \rho_{gas})}{\mu_w} * d_{ab}^2 \qquad (d_{ab} < 0.1 \text{ mm}) \quad (C\text{-}17)$$

For small rigid spherical bubbles (i.e. $0.1 < d_{ab} < 1$ mm), the rise velocity is best fitted by (COMOLET 1979a) :

$$u_r = \frac{g * \rho_w}{18 * \mu_w} * d_{ab}^2 \qquad (0.1 < d_{ab} < 1 \text{ mm}) \quad (C\text{-}18)$$

As far as the fluid viscosity is neglected (i.e. $d_{ab} > 1$ mm) COMOLET (1979b) showed that the bubble rise velocity can be estimated as :

$$u_r = \sqrt{\frac{2.14 * \sigma}{\rho_w * d_{ab}} + 0.52 * g * d_{ab}} \qquad (d_{ab} > 1 \text{ mm}) \quad (C\text{-}19)$$

Figure C-2 summarises these results

C.5 Acknowledgment

The present methodology was developed by the author who did also the calculations. Ms QIAO G.L. did separate calculations as part of her Ph.D. project.

Fig. C-2 - Bubble rise velocity in still water - COMOLET (1979a)

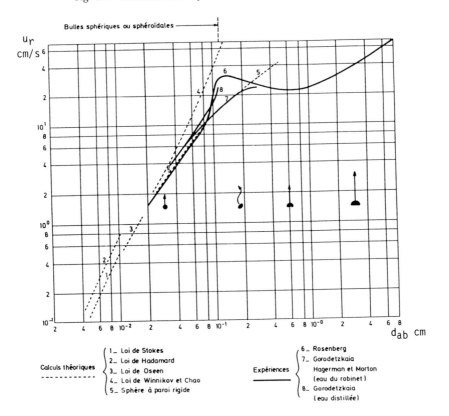

	1_ Loi de Stokes		6_ Rosenberg
Calculs théoriques	2_ Loi de Hadamard	Expériences	7_ Gorodetzkaia
-------------	3_ Loi de Oseen		Hagerman et Morton
	4_ Loi de Winnikov et Chao		(eau du robinet)
	5_ Sphère à paroi rigide		8_ Gorodetzkaia
			(eau distillée)

APPENDIX D

MODELLING AIR-WATER GAS TRANSFER IN SKIMMING FLOWS

D.1 Introduction

Fick's law states that the mass transfer rate of a chemical across an interface varies directly as the coefficient of molecular diffusion D_{gas} and the gradient of gas concentration. For the gas transfer of a dissolved gas across an air-water interface, it is usual to rewrite Fick's law as :

$$\frac{d}{dt} C_{gas} = K_L * a * (C_s - C_{gas}) \tag{D-1}$$

where C_{gas} is the concentration of dissolved gas, K_L is the coefficient of mass transfer, a is the specific surface area defined as the air-water interface area per unit volume of air and water and C_s is the gas saturation concentration. At each position (x,y) along a streamline, equation (D-1) can be averaged over a small control volume and it yields :

$$\frac{d}{ds} C_{gas}(x,y) = \frac{K_L(x,y) * a(x,y)}{V(x,y)} * (C_s(x,y) - C_{gas}(x,y)) \tag{D-2}$$

where V is the local flow velocity and s is the direction along the streamline.

In free-surface flows along a chute, the coefficient of transfer K_L and the saturation concentration C_s can be assumed constant. Along the chute and at each location, equation (D-2) can be averaged over the cross-section. It yields (CHANSON 1993d) :

$$\frac{d}{ds} C_{gas} = \frac{K_L * a_{mean}}{U_w} * (C_s - C_{gas}) \tag{D-3}$$

where U_w is the mean flow velocity and a_{mean} is the mean specific interface area.

D.2 Hydraulic characteristics

In self-aerated flows, the air concentration distribution can be estimated by a diffusion model of the air bubbles within the air-water mixture (WOOD 1984) :

$$C = \frac{B'}{B' + \exp\left(-G'^* \cos\alpha * \left(\frac{y}{Y_{90}}\right)^2\right)} \tag{D-4}$$

where C is the local air concentration, B' and G' are functions of the mean air concentration only (table D-1), α is the spillway slope, y is the distance measured perpendicular to the invert and Y_{90} is the depth where the air concentration is 90%.

Next to the invert, experimental data depart from equation (D-4) and the air concentration tends to zero at the bottom (CHANSON 1992b). The data shows consistently the presence of an air concentration boundary layer in which the air concentration distribution is :

$$\frac{C}{C_b} = \left(\frac{y}{\delta_{ab}}\right)^{0.270} \tag{D-5}$$

Table D-1 - Air concentration distribution coefficients (after CHANSON 1992a, 1993b)

C_{mean}	$G'^*\cos\alpha$ [a]	B' [a]	C_b [b]	$\dfrac{V_{90}{}^*Y_{90}}{q_w}$
(1)	(2)	(3)	(4)	(5)
0.0	+ infinite	0	0	1.167
0.1608	7.999	0.003021	0.02	1.453
0.2411	5.744	0.028798	0.04	1.641
0.3100	4.834	0.07157	0.07	1.805
0.4104	3.825	0.19635	0.17	2.141
0.5693	2.675	0.62026	0.36	2.985
0.6222	2.401	0.8157	0.46	3.319
0.6799	1.8942	1.3539	0.55	4.151
0.7209	1.5744	1.8641	0.64	4.859

Notes : [a]   computed from STRAUB and ANDERSON's (1958) data
     [b]   computed from equation (D-4) : $C_b = B'/(B'+1)$

where C_b is the air concentration at the outer edge of the air concentration boundary layer and δ_{ab} is the air concentration boundary layer thickness. On Aviemore spillway, the author estimated δ_{ab} = 15 mm. C_b satisfies the continuity between equations (D-4) and (D-5). For spillway flows, the characteristic depth Y_{90} is much larger than δ_{ab} and a reasonable approximation is : $C_b \sim B'/(1+B')$ (table D-1).

Velocity measurements in self-aerated flow showed that the velocity distribution follows a power law distribution :

$$\frac{V}{V_{90}} = \left(\frac{y}{Y_{90}}\right)^{1/N} \tag{D-6}$$

Note that equation (D-6) is independent of the mean air concentration. For a given mean air concentration, the characteristic velocity V_{90} is deduced from the continuity equation for the water flow (table D-1). On stepped chutes, the exponent N can be deduced from experimental measurements (FRIZELL 1992, TOZZI 1992) or from equation (4-39) : i.e. $N = K^*\sqrt{8/f}$ where K is the Von Karman constant (K = 0.40) and f is the friction factor.

D.3 Air-water interface area

In a turbulent shear flow, the maximum air bubble size is determined by the balance between the capillary and inertial forces. The author (CHANSON 1992a, 1993d) showed that the maximum bubble size can be estimated as :

$$d_m \sim \sqrt[3]{\frac{2 * \sigma * (We)_c}{\rho_w * \left(\dfrac{dV}{dy}\right)^2}} \tag{D-7}$$

where the velocity gradient dV/dy is deduced from equation (D-6). Experiments have shown that the critical Weber number $(We)_c$ is a constant near unity (table 6-7). With the hypothesis that : {H1} the critical Weber number equals unity and {H2} the bubble diameter is in order of magnitude of the mixing length, equation (D-7) provides an estimate of the maximum bubble size in self-aerated skimming flows.

High speed photographs showed that the shape of air bubbles in self-aerated flows is approximately spherical and the specific interface area equals :

$$a = 6 * \frac{C}{d_{ab}} \tag{D-8a}$$

for air bubbles in water. For water droplets in air, the interface area is :

$$a = 6 * \frac{1 - C}{d_{wp}} \tag{D-8b}$$

where d_{wp} the diameter of water particles. The calculation of the air-water interface requires the definition an "ideal" air-water interface. An advantageous choice is the location where the air concentration is 50%. This definition is consistent with experimental observations (CHANSON 1993d) and satisfies the continuity of equation (D-8).

Assuming {$d_{ab} \sim d_m$}, equations (D-7) and (D-8) enable the calculation of the air-water surface area in the air-in-water flow. Little information is available on the size of the water droplets surrounded by water. Photographs suggest that the droplet sizes are of similar size as the large air bubbles (VOLKART 1980) and equation (D-7) may provide a first estimate of the droplet sizes. It must be emphasised that the contribution of large particles (bubbles or droplets) to the interface area is relatively small.

D.4 Aeration efficiency of stepped chutes with skimming flow

Downstream of the inception point, the author developed a simple numerical model to compute the aerated flow characteristics on smooth spillway (CHANSON 1993a). In the chapter 4, some similarities between free-surface aerated flows on smooth chutes and stepped chute flows with skimming flow regime have been shown. Using this analogy, the air-water gas transfer of skimming flows on stepped chutes can be estimated using the method of CHANSON (1993d) downstream of the inception point of air entrainment (paragraph 4.3).

At any position along the spillway, the distributions of (un-dissolved) air concentration, velocity, bubble size and air-water interface area can be estimated using equations (D-4) & (D-5), (D-6), (D-7) and (D-8) respectively. The air-water gas transfer can be deduced from equation (D-3) where the mean interface area equals :

$$a_{mean} = \frac{1}{Y_{90}} * \int_{y=0}^{y=Y_{90}} a * dy \tag{D-9}$$

APPENDIX E
UNIT CONVERSIONS

E.1 Introduction

The systems of units derived from the metric system have gradually given way to a single system, called the Système International d'Unités (SI) and the present monograph presents results expressed in SI Units.

Since a large number of countries continue to use British and American units, this appendix gives their equivalents against the SI units.

References : DEGREMONT (1979), International Organization for Standardization (1979)

E.3 Units and conversion factors

Quantity (1)	Unit (symbol) (2)	Conversion (3)	Comments (4)
Length :	1 inch (in)	$= 25.4 \ 10^{-3}$ m	Exactly.
	1 foot (ft)	$= 0.3048$ m	Exactly.
	1 yard (yd)	$= 0.9144$ m	Exactly.
	1 mile	$= 1.609.344$ m	Exactly.
Area :	1 square inch (in^2)	$= 6.4516 \ 10^{-4} \ m^2$	Exactly.
	1 square foot (ft^2)	$= 0.09290306 \ m^2$	Exactly.
Volume :	1 cubic inch (in^3)	$= 16.387064 \ 10^{-6} \ m^3$	Exactly.
	1 cubic foot (ft^3)	$= 28.3168 \ 10^{-3} \ m^3$	Exactly.
	1 gallon UK (gal UK)	$= 4.54609 \ 10^{-3} \ m^3$	
	1 gallon US (gal US)	$= 3.78541 \ 10^{-3} \ m^3$	
	1 barrel US	$= 158.987 \ 10^{-3} \ m^3$	Petroleum,....
Velocity :	1 foot per second (ft/s)	$= 0.3048$ m/s	Exactly.
	1 mile per hour (mph)	$= 0.44704$ m/s	Exactly.
Acceleration :	1 foot per second squared (ft/s^2)	$= 0.3048 \ m/s^2$	Exactly.
Mass :	1 pound (lb or lbm)	$= 0.45359237$ kg	Exactly.
	1 ton UK	$= 1016.05$ kg	
	1 ton US	$= 907.185$ kg	
Density :	1 pound per cubic foot (lb/ft^3)	$= 16.0185 \ kg/m^3$	
Force :	1 kilogram-force (kgf)	$= 9.80665$ N (exactly)	Exactly.
	1 pound force (lbf)	$= 4.4482216152605$ N	
Moment of force :	1 foot pound force (ft.lbf)	$= 1.35582$ N.m	
Pressure :	1 Pascal (Pa)	$= 1 \ N/m^2$	
	1 standard atmosphere (atm)	$= 101325$ Pa	Exactly.
	1 bar	$= 10^5$ Pa	Exactly.
	1 Torr	$= 133.322$ Pa	
	1 conventional metre of water (m of H_2O)	$= 9.80665 \ 10^3$ Pa	Exactly.
	1 conventional meter of Mercury (m of Hg)	$= 1.33322 \ 10^5$ Pa	
	1 Pound per Square Inch (PSI)	$= 6.8947572 \ 10^3$ Pa	
Temperature :	T (Celsius)	$= $ T (Kelvin) $- 273.16$	0 Celsius is 0.01 K below the temperature of the triple point of water.
	T (Fahrenheit)	$= T \ (Celsius) * \dfrac{9}{5} + 32$	
	T (Rankine)	$= \dfrac{9}{5} * T \ (Kelvin)$	
Dynamic viscosity :	1 Pa.s	$= 0.006720$ lbm/ft/s	
	1 Pa.s	$= 10$ Poises	
Kinematic viscosity :	1 square foot per second (ft^2/s)	$= 0.0929030 \ m^2/s$	
	1 m^2/s	$= 10.7639 \ ft^2/s$	
	1 m^2/s	$= 10^4$ Stokes	
Work energy :	1 Joule (J)	$= 1$ N.m	
	1 Joule (J)	$= 1$ W.s	
	1 Watt hour (W.h)	$= 3.600 \ 10^3$ J	Exactly.
	1 electronvolt (eV)	$= 1.60219 \ 10^{-19}$ J	
	1 foot pound force (ft.lbf)	$= 1.35582$ J	
Power :	1 Watt (W)	$= 1$ J/s	
	1 foot pound force per second (ft.lbf/s)	$= 1.35582$ W	
	1 horsepower (hp)	$= 745.700$ W	

APPENDIX F
CORRECTIONS

F.1 Introduction

If you find a mistake or an error, your help and assistance would be most appreciated to improve the next edition of this monograph. Thank you to record the errors on this page. Please make a copy of it and send it to the author :

Dr H. CHANSON

Dept. of Civil Engineering, The University of Queensland, Brisbane QLD 4072, Australia

Fax.:   (61 7) 365 45 99   Email : chanson@uq_civil.civil.uq.oz.au

F.2 Correction form

Contact

Name :   ....

Address :   ....

....

....

Tel.:   ....

Fax: ....

Email :   ....

Description of the error

Page number :   Line number :   Equation number :

....

....

....

....

....

Proposed correction

....

....

....

....

....

....

INDEX

Note :   (*f*)   figure or photograph